T0073467

Jeffrey W. Tweedale and Lakhmi C. Jain

Embedded Automation in Human-Agent Environment

# Adaptation, Learning, and Optimization, Volume 10

## Series Editor-in-Chief

Meng-Hiot Lim
Nanyang Technological University, Singapore
E-mail: emhlim@ntu.edu.sg

Yew-Soon Ong
Nanyang Technological University, Singapore
E-mail: asysong@ntu.edu.sg

Further volumes of this series can be found on our homepage: springer.com

Vol. 1. Jingqiao Zhang and Arthur C. Sanderson
Adaptive Differential Evolution, 2009
ISBN 978-3-642-01526-7

Vol. 2. Yoel Tenne and Chi-Keong Goh (Eds.)
Computational Intelligence in
Expensive Optimization Problems, 2010
ISBN 978-3-642-10700-9

Vol. 3. Ying-ping Chen (Ed.)
Exploitation of Linkage Learning in Evolutionary Algorithms, 2010
ISBN 978-3-642-12833-2

Vol. 4. Anyong Qing and Ching Kwang Lee
Differential Evolution in Electromagnetics, 2010
ISBN 978-3-642-12868-4

Vol. 5. Ruhul A. Sarker and Tapabrata Ray (Eds.)
Agent-Based Evolutionary Search, 2010
ISBN 978-3-642-13424-1

Vol. 6. John Seiffertt and Donald C. Wunsch
Unified Computational Intelligence for Complex Systems, 2010
ISBN 978-3-642-03179-3

Vol. 7. Yoel Tenne and Chi-Keong Goh (Eds.)
Computational Intelligence in Optimization, 2010
ISBN 978-3-642-12774-8

Vol. 8. Bijaya Ketan Panigrahi, Yuhui Shi, and Meng-Hiot Lim (Eds.)
Handbook of Swarm Intelligence, 2011
ISBN 978-3-642-17389-9

Vol. 9. Lijuan Li and Feng Liu
Group Search Optimization for Applications in Structural Design, 2011
ISBN 978-3-642-20535-4

Vol. 10. Jeffrey W. Tweedale and Lakhmi C. Jain
Embedded Automation in Human-Agent Environment, 2011
ISBN 978-3-642-22675-5

Jeffrey W. Tweedale and Lakhmi C. Jain

# Embedded Automation in Human-Agent Environment

 Springer

Dr. Jeffrey W. Tweedale
University of South Australia
School of Electrical and Information
Engineering
Mawson Lakes Campus
Adelaide
South Australia SA 5095
Australia
E-mail: Jeff.Tweedale@unisa.edu.au

Dr. Lakhmi C. Jain
University of South Australia
School of Electrical and Information
Engineering
Mawson Lakes Campus
Adelaide
South Australia SA 5095
Australia
E-mail: Lakhmi.jain@unisa.edu.au

ISBN 978-3-642-22675-5                    e-ISBN 978-3-642-22676-2

DOI 10.1007/978-3-642-22676-2

Adaptation, Learning, and Optimization        ISSN 1867-4534

Library of Congress Control Number: 2011934494

*Typeset* & *Cover Design:* Scientific Publishing Services Pvt. Ltd., Chennai, India.

Printed on acid-free paper

9 8 7 6 5 4 3 2 1

springer.com

*To everyone that helped!*

# Preface

In the last few years, we have witnessed a tremendous growth in the applications of multi-agent paradigms in every discipline right from health sciences, engineering, management, to aviation. This has also created tremendous interests among researchers in the architecture, learning, communication, cooperation, teaming and automation of multi-agent systems. The advances in automation have enabled the formation of rather unique teams between agents and human and between agents and agents.

The book proposes a general conceptual framework for the development of automation in human-agents environments that will allow such teams to work effectively and efficiently. The book includes twelve chapters. Chapter 1 provides an introduction to embedded automation. Chapter 2 describes the innovations driving modern artificial intelligence and agent teaming. Chapter 3 presents a sample of literature that briefly reviews automation using multi-agent systems. Chapter 4 presents the evolution of agent technology and frameworks to the present day design.

Chapter 5 is on agent interoperability, adaptation and communication. Chapter 6 is on enhancing autonomy while Chapter 7 is on improving agent architectures (from blackboards to micro-simulation with forward synthesis). Chapter 8 presents a number of frameworks and agent implementation techniques. Chapter 9 is on agent oriented programming concepts useful for the successful implementation of automation using multi-agent systems. Chapter 10 describes the strategies and techniques currently used by humans to solve problems. Chapter 11 presents the case study used to illustrate the autonomous methods used to solve real world problems and focuses on the Sudoku puzzle. The final chapter (12) presents the concluding remarks and comments on the future directions.

Many scientists, application engineers, academic professionals and students will find this book useful when considering agency solutions within their problem domain.

We both wish to thank our families who have been so patient with our late nights and their support in writing this book. Jeffrey wishes to express his gratitude to his supervisors at the Defence Science and Technology Organisation for their trust and support while rationalising the content. He would also like to express his gratitude to those who saw him through this experience, especially his colleagues within the Airborne Missions System branch and at the University who provided support, talked things over and offered comments.

We appreciate the constructive comments provided by the reviewers which helped in improving the clarity of presentation of key concepts. We also believe that there is sufficient academic and commercial support for agents. They are increasing being used to solve progressively more complex problems. As we approach applications that solve real-world problems, the skills required have risen dramatically. Practical examples of where agents are currently used, include: spell checking, spam filters, travel and event booking systems.

We acknowledge the editorial assistance by Springer-Verlag.


Dr. Jeffrey W. Tweedale
Adelaide, Australia


Dr. Lakhmi C. Jain
Adelaide, Australia

# Contents

# Acronyms

| | |
|---|---|
| **ACCESS** | Agents Channelling ContExt Sensitive Services |
| **ACL** | Agent Communication Languages |
| **Æ** | Arithmetic Expression |
| **AFC** | Application Foundation Classes |
| **AFD** | Agent Factory Demonstrator |
| **AFP** | Advanced Flexible Processor |
| **AFAPL** | Agent Factory Agent Programming Language |
| **AFSE** | Agent Factory Standard Edition |
| **AIP** | Advanced Information Processing |
| **AI** | Artificial Intelligence |
| **ALU** | Arithmetic Logic Unit |
| **AMD** | Advanced Micro Devices |
| **AM** | Attention Management |
| **ANA** | Agent Network Architecture |
| **ANN** | Artificial Neural Network |
| **AOP** | Aspect Oriented Programming |
| **AOPL** | Agent Oriented Programming Language |
| **AOS** | Agent Oriented Software |
| **APB** | Advanced Processing Build |
| **API** | Application Program Interface |
| **ASF** | Apache Software Foundation |
| **ASIC** | Application-Specific Integrated Circuit |
| **ASM** | refers to the asm keyword |
| **ATLAS** | Agent Transportation Layer Adaption System |
| **ATMS** | Assumption-based Truth Maintenance System |
| **AWT** | Abstract Windowing Toolkit |
| **BCEL** | Byte Code Engineering Library |
| **BDI** | Beliefs, Desires, Intentions |
| **BNF** | Backas-Normal Form |
| **BPEL** | Business Process Execution Language |
| **BPM** | Business Process Model |

| | |
|---|---|
| **C²** | Command and Control |
| **CAST** | Collaborative Agent for Simulating Teamwork |
| **CBR** | Cased-Based Reasoning |
| **CDC** | Control Data Corporation |
| **CDI** | Java Contexts and Dependency Injection for the Java EE platform |
| **CE** | Copilote Electronique |
| **CFG** | Context-Free Grammar |
| **CFL** | Context-Free Language |
| **CIA** | Communication Interchange Agent |
| **CI** | Computational Intelligence |
| **CMOS** | Complimentary Metal Oxide Semiconductor |
| **COAK** | Capture Operator or Artisan Knowledge |
| **COA** | Course Of Action |
| **Corba** | Common Object Request Broker Architecture |
| **CPU** | Central Processing Unit |
| **CRM** | Cockpit Resource Management |
| **CSP** | Constriaint Programming |
| **DAI** | Distributed Artificial Intelligence |
| **DAML-S** | DARPA Agent Markup Language - Services |
| **DAML** | DARPA Agent Markup Language |
| **DARPA** | Defense Advanced Research Projects Agency |
| **DC-AOP** | Dynamic Composition for Autonomous Object Platforms |
| **DCE** | Distributed Computing Environment |
| **DCF** | Dynamic Composition of Functionality |
| **DCOM** | Distributed Component Object Model |
| **DHTML** | Dynamic HTML |
| **DIARG** | Dynamic Inter-Agent Rule Generation |
| **DIS** | Distributed Intelligent Systems |
| **dMARS** | distributed Multi-Agent Reasoning System |
| **DPS** | Distributed Problem Solving |
| **DS1** | Deep Space One |
| **DSP** | Digital Signal Processor |
| **DSS** | Decision Support System |
| **DSTO** | Defence Science and Technology Organisation |
| **DTD** | Document Type Definition |
| **EH** | Exception Handling |
| **EJB** | Enterprise Java Beans |
| **EMA** | Ends Means Analysis |
| **EMMA** | Enterprise Management and Modelling Architecture |
| **FA** | Finite Automation |
| **FIPA ACL** | FIPA Agent Communication Languages |
| **FIPA** | Foundation of Intelligent Physical Agents |
| **FLC** | Fuzzy Logic Control |
| **FOPL** | First Order Predicate Logic |
| **FPGA** | Field Programmable Grid or Gate Arrays |

| **FRAM** | Ferromagnetic Random Access Memory |
| **FraMAS** | Framework for Multi-Agent Systems |
| **FTP** | File Transfer Protocol |
| **GA** | Genetic Algorithm |
| **GCE** | Generalized Constraint Enforcement |
| **GD** | Generalized Damping |
| **GoF** | Gang of Four |
| **Golog** | alGOl Logical prOGramming |
| **GPU** | Graphics Processing Unit |
| **GUI** | Graphical User Interface |
| **HCA** | Human Centered Automation |
| **HCI** | Human Computer Interface |
| **HCT** | Human-Computer Trust |
| **HDD** | Hard Disk Drive |
| **HEPE** | High-Energy Physics Experiments |
| **HIL** | human-in-the-loop |
| **HMI** | Human Machine Interface |
| **HPC** | High Performance Computing |
| **HTML** | HyperText Markup Language |
| **HTTP** | HyperText Transfer Protocol |
| **IA** | Intelligent Agent |
| **IBM** | International Business Machines |
| **IC** | Integrated Circuit |
| **IDE** | Integrated Development Environment |
| **IDL** | Interface Description Language |
| **IDSS** | Intelligent Decision Support System |
| **IIOP** | Internet inter-ORB Protocol |
| **IMM** | Individual Mental Model |
| **InteRRaP** | Integration of Reactive Behavior and Rational Planning |
| **IOI** | Input-Output Maps using Interpolation |
| **I/O** | Input/Output |
| **OSIRM** | Open Systems Interconnection Reference Model |
| **ISO** | International Standards Organisation |
| **ISS** | International Space Station |
| **ITMS** | Incremental Truth Management System |
| **JAAS** | Java Authentication and Authorisation Service |
| **JACK** | Java Agent Compiler and Kernel |
| **JADE** | Java Agent Development Environment |
| **JAF** | JavaBeans Activation Framework |
| **JARE** | Java Automatic Reasoning Engine |
| **JAXP** | Java API for XML Processing |
| **JAX-RPC** | Java API for XML based RPC |
| **JAX-WS** | Java XMLWeb Services |
| **JAXM** | Java API XML Messaging |
| **JDBC** | Java Data Base Connectivity |

| | |
|---|---|
| **JDIC** | JDesktop Integration Components |
| **JDK** | Java Developers Kit |
| **JIT** | Just-in-Time |
| **JMS** | Java Messaging System |
| **JMX** | Java Management Extension |
| **JNDI** | Java Naming and Directory Interface |
| **JNI** | Java Native Interface |
| **JPDA** | Java Platform Debugger Architecture |
| **JPL** | Jet Propulsion Laboratories |
| **JPS** | Java virtual machine Process Status tool |
| **JRE** | Java Run-time Environment |
| **JSP** | JavaServer Pages |
| **JSR** | Java Specification Program |
| **JS** | Java Script |
| **JVM** | Java Virtual Machine |
| **KBS** | Knowledge Based System |
| **KES** | Knowledge-Based Intelligent Information and Engineering Systems |
| **KIF** | Knowledge Interchange Format |
| **KQML** | Knowledge Query Manipulation Language |
| **KSE** | Knowledge Sharing Effort |
| **LA** | Local Adaptation |
| **LCD** | Liquid Crystal Display |
| **LEAP** | Lightweight Extensible Agent Platform |
| **LIFO** | Last In, First Out |
| **LISP** | LISt Processing |
| **LSI** | Large Scale Integration |
| **LTMS** | Logical Truth Maintenance System |
| **MALLET** | Multi-Agent Logic Language for Encoding Network |
| **MAF** | Mobile Agent Facility |
| **MAS** | Multi-Agent System |
| **MASIF** | Mobile Agent System Interoperability Facility |
| **MBC** | Model-Based Computing |
| **MFG** | Membership Function Generator |
| **MLC** | Multi-Level Cell |
| **MMA** | Mission Management Aid |
| **MOE** | Measures Of Efficiency |
| **MOP** | Measures Of Performance |
| **MPF** | Mars Pathfinder |
| **MP** | Multi-Processor |
| **MT** | Multi-Threaded |
| **MURI** | Multidisciplinary Research Program of the University Research Initiative |
| **MVC** | Model View Control |
| **NAND** | NOT AND |
| **NASA** | National Aeronautics and Space Administration |

| | |
|---|---|
| **NCR** | National Research Council |
| **NewMAAP** | New Millennium Autonomy Architecture Prototype |
| **NIO** | New I/O |
| **NP** | Non-deterministic Polynomial |
| **OMG** | Object Management Group |
| **ONR** | Office of Naval Research |
| **OODA** | Observe Orient Decide and Act |
| **OOL** | Object Oriented Languages |
| **OOPL** | Object Oriented Programming Language |
| **OOPS** | Object Oriented Programming Software |
| **OOP** | Object Oriented Programming |
| **ORB** | Object Request Broker |
| **OSIRM** | Open Systems Interconnection Reference Model |
| **OSI** | Open Systems Interconnection |
| **OS** | Operating System |
| **OWL-S** | OWL-Services |
| **PAL** | Programmable Array Logic |
| **PA** | Pilot's Associate |
| **PDA** | Push Down Autonoma |
| **PID** | Process IDentifier |
| **PDA** | Personal Digital Assistant |
| **PM** | Process Management |
| **PMM** | Process Manager Module |
| **PNML** | Petri-Net Markup Language |
| **PnP** | Plug and Play |
| **POJO** | Plain Old Java Object |
| **POSIX** | Portable Operating System Interface |
| **PROLOG** | PROcedural LOGic |
| **PRS** | Procedural Reasoning System |
| **R-CAST** | Recognition-Primed Collaborative Agent for Simulating Teamwork |
| **RAM** | Random Access Memory |
| **REST** | Representational State Transfer |
| **RISC** | Reduced Instruction Set Computer |
| **RMI** | Remote Method Invocation |
| **RPC** | Remote Procedure Call |
| **RPD** | Recognition-Primed Decision |
| **SDK** | Software Development Kit |
| **SGML** | Standard Generalized Markup Language |
| **SME** | Subject Matter Expert |
| **SMM** | Shared Mental Model |
| **SMP** | Symmetric Multi-Processor |
| **SNMP** | Simple Network Management Protocol |
| **SOAP** | Simple Object Access Protocol |
| **SOAR** | State, Operator And Result |
| **SOA** | Service-Oriented Architecture |

| | |
|---|---|
| **SoC** | System on Chip |
| **SODA** | Stimulate Observe Decide and Act |
| **SQL** | Structured Query Language |
| **SRK** | Skills, Rules and Knowledge |
| **SSD** | Solid State Drives |
| **SST** | Situation Specific Trust |
| **TCP/IP** | Transmission Control Protocol / Internet Protocol |
| **TMS** | Truth Maintenance System |
| **TM** | Turing Machine |
| **TNC** | Trust, Negotiation, Communication |
| **TSP** | Traveling Salesman Problem |
| **UAS** | Unmanned Aerial System |
| **UAV** | Unmanned Aerial Vehicle |
| **UDP** | User Datagram Protocol |
| **UGV** | Unmanned Ground Vehicle |
| **UI** | User Interface |
| **UML** | Unified Modelling Language |
| **UPnP** | Universal Plug and Play |
| **URL** | Universal Resource Locator |
| **UUV** | Underwater Unmanned Vehicle |
| **VHDL** | Very High Speed Integrated Circuit Hardware Description Language |
| **VLSI** | Very Large Scale Integration |
| **WAIS** | Wechsler Adult Intelligence Scale |
| **WFF** | Well Formed Formula |
| **WSDL** | Web-Services Description Language |
| **WWW** | World Wide Web |
| **XHTML** | Extensible HyperText Markup Language |
| **XML** | Extensible Markup Language |
| **XOM** | XML Object Model |

*"Never abandon a theory that explains something until you have a theory that explains more [245]."*

John McCarthy

# 1

# Introduction to Embedded Automation

## 1.1 Introduction

This book has resulted from ongoing research by a group of enthusiastic people within both Defence Science and Technology Organisation (DSTO) and Knowledge-Based Intelligent Information and Engineering Systems (KES) Centre during work on automation and Artificial Intelligence (AI). Our effort has enabled the formation of teams that combine the skills of both human and machine (electronic) members. This book proposes a general, conceptual framework for the development of automation in human-machine environments that will allow such teams to work effectively and efficiently. This chapter introduces the subject and the main challenges which developers have to face. The method of approach in developing this conceptual framework will be described with an outline of original contributions. An overview of the structure of the book and its chapters is also presented.

## 1.2 Challenges

The book provides a viable mechanism to enable agents to auto-negotiate a communications channel, using self organising techniques to establish trust via negotiation in order to share the system workload or maintain a balanced and efficient application. The use of multi-agent teams is also becoming entrenched in applications that requires significant Human Computer Interface (HCI). In the Air domain a hierarchy of agents could be used to coordinate and control a team of Unmanned Aerial Vehicle (UAV) or coordinate a squadron of aircraft during a mission. Agents in the form of an adaptive HCI may be used to assist in improving the efficiency of

J.W. Tweedale & L.C. Jain: Embedded Automation in Human-Agent Environment, ALO 10, pp. 1–14.
springerlink.com

team-based applications. Hence, the Defence community is very interested in the progress of this research and its possible uses within their equipment.

Therefore this study will:

- Identify the architecture and implementation of Multi-Agent System (MAS) that could be used to improve coordination and control using the Trust, Negotiation, Communication (TNC) concepts;
- Identify the properties required to negotiate a trusted role as a team member capable of being controlled or coordinated from a variety of sources; and
- Examine the adaptable nature of agent teams, to dynamically reconfigure their capability and solve problems.

## 1.3   Approach

This book is intended for an audience of scientists, students, researchers and application engineers who are interested in the analysis, design and development of automation applications in complex adaptive environments. The main contribution of this book is the definition of a coherent, conceptual agency framework for automation and a clear articulation of generic steps to be taken in this process. These steps are:

Definition of Boundaries:  The basis of the framework will be the introduction of automation (machine-assistant) into the traditional environment where the human is ultimately responsible for all activities. A classical system engineering approach will be used to identify the boundaries, interfaces and tasks to be shared by human and machine. The nature of the relationship between adaptation and automation will also be characterised.

Task Classification:  The classification of tasks can be allocated to either the human or machine. Two main schemes for task classification, will be used, based on the interface and the level of control required. These are not mutually exclusive, but each should be capable of dynamic interaction and able to self organize in a coherent manner to adapt within the environment.

Task Management and Coordination:  The use of automated coordination and collaboration to achieve the successful completion of tasks is a requirement introduced by automation. This provides a mechanism for MAS to supervise and cooperate on tasks to establish the extended trust and adaptability of the agent system during evolution.

## 1.4   Outline of Book

The introduction reflects the problem domain, its challenges and the method of approach used to develop in order to examine the requirements to extend the self organizing agents, teams or systems. The structure used is designed to develop the integrity of the topic and maintain the readability of each subject. Although it is difficult to write stand-alone chapters without duplicating text, over-use of this practice has been avoided where possible. Beginning in chapters 2, 3 and 4, the reader is guided through the relevant background of problems associated with automation using MAS, complete with a description of the tools, technology and training that has evolved, with examples of existing research that support agents, teams and AI. Chapter 2 describes the innovations currently available on modern artificial intelligence. Due the the spans of time, a significant number of references have been used. Many of the older references are books by Subject Matter Experts (SMEs) or hard to find articles, while more of the current references are from within the research community. The background of AI[1] complete with a brief history to help put many of the pioneering milestones into perspective are covered. This covers the basic concepts of logic, mathematical tools, aids and programming techniques, through to the pioneering milestones surrounding the development of the Central Processing Unit (CPU) and the modern day computer. The major discussion revolves around agents, MAS and how Teams can be used to embed intelligence into mainstream applications. Chapter 3 reviews existing research in automation using AI techniques, how knowledge is represented, stored and decisions made. Especially important is the role of a human in any system, the problems they bring and assumptions used in order to minimize this effect. Chapter 4 describes the evolution of computer technology contributing to the field of AI. The types of architecture, existing research, limitations and the authors method of moving to a dynamic capability within an integrated system.

Communications, interoperability and information exchange are tackled in chapter 5. The definition of syntax, semantics, topology and ontology are laid-out to ensure the lexicon between the reader and the author remain synchronized. A large portion of the chapter is consumed with the evolution of existing communication languages from Agent Communication Languages (ACL) to Simple Object Access Protocol (SOAP). Additional web-service protocols follow to complete the review. The principles raised explain how each has their own use, but the facts remain that no single language is suitable to conduct a meaningful and effective exchange of data, information and knowledge, required either at the physical, transmission or cognitive levels.

Autonomy is a major feature of this research and in chapter 6. Significant material is offered to display the amount of effort expended on existing research. We cover components, design patterns, threads and there relationships. We discuss several research topics concerning interoperability and trust. A number of the experiments conducted are used to explain how the derivation of the problem is achieved. These include the Agent Transportation Layer Adaption System (ATLAS) and Trust, Negotiation, Communication (TNC) models, as well as a concept demonstrator used to enhance the dynamic nature of teaming within a distributed application.

---

[1] McCarthy has stated he would have preferred to use the term Computational Intelligence (CI).

The scale and complexity of agent systems has escalated. They now demand significant programming knowledge and experience, with specialty skills in agency theory and practice. Chapter 7 discusses a number of improvements made to agent frameworks and structural architectures. We begin with advances in silicon abstraction using an Field Programmable Grid or Gate Arrays (FPGA), discussing fuzzy processors, Cased-Based Reasoning (CBR) systems, Knowledge Based System (KBS), Artificial Neural Network (ANN) and the concept of dynamically reconfiguring Very Large Scale Integration (VLSI) coprocessors. We follow with theory about trust, Service-Oriented Architecture (SOA), dynamic architectures and possible micro-simulation that incorporates problem synthesis (forward-chaining scenarios).

Agent oriented programming is becoming a popular concept. We currently have a number of enterprise level scripting languages that assist in providing autonomous task processing. Chapter 8 presents a brief history of agent development, existing frameworks and implementation techniques. We provide a progressive discussion about spiders, bots and aggregators, before exploring agents. To explain agent mobility, we discuss connectivity, data exchange (passing) and new concepts on how to structure or decompose problem solving. We also introduce the concept of using patterns, proxies and generics.

Chapter 9 is on agent oriented programming concepts useful for the successful implementation of automation using multi-agent systems. This chapter discusses the concepts surrounding agent oriented programming that are useful for the successful implementation of automation using multi-agent systems. We use the Java language and discuss recent enhancements to its current libraries and any proposed changes. Topics include MAS frameworks, transformations and how programs are instantiated and accessed in memory. We also describe how you can now dynamically install agents using proxies, dynamic proxies and how these techniques can be used to automate class loading during run-time. There is also a brief discussion about basic structure and suggested functionality required in an *agent factory*.

AI researchers traditionally focus on game related problems in order to demonstrate a specific technique or solution. Chapter 10 defines the problem relating to sudoku, which requires a number of cognitive techniques to successfully solve each puzzle. We explain the strategies and techniques commonly used by humans to solve the problem. Many of these techniques are demonstrated using a practical walk through.

To achieve any of the enhancements suggested, a paradigm shift must occur. The current process of enhancing performance using additional hardware needs to change. Future enhancements can, and should, be made *within* the frameworks and architectures supporting an application. In chapter 11, a number of techniques are discussed that could be enhanced to improve the efficiency of many AI techniques currently run on computers. A case study is used to illustrate the autonomous methods used to solve real world problems and focuses on the Sudoku puzzle. The use of dynamic programming, open architectures, polymorphism, Advanced Information Processing (AIP), distributed systems and enhanced CPU design need to be revised in order to aim at applications capable of real-time and faster than real-time execution. Here the concept of micro-simulation capability is raised and its use has the potential to substantially improve science in many areas.

Future comments and concluding remarks are contained in chapter 12. They include a wrap-up of proposed software paradigms, possible agent interoperability architectures or frameworks, the change required in technology and microprocessor architectures, a new methodology for automation, adaption and function switching. The basic premise illustrates that although technology has not grown to meet the needs of the AI world, avenues exist that will improve the efficiency of the techniques used to solve problems by tailoring one or more existing methodology.

## 1.5 Way Forward

The information age [182] has introduced many changes to the way we work, rest and play. The *tools*, *technology* and *training* that have been developed during the past century are truly amazing. In this context, tools provide an interface between two entities that assist one domain within another, while technology has been a constant source of innovation in modern science or academic pursuit and should be considered as a collection of techniques employed in a systematic approach[2]. On the other hand where training is used to increase competency (gained through the transfer of rules, skills or knowledge) in the community that employees these concepts. All three terms form the basis of a variety of themes that can be used to describe the what, who, where, when and how properties relating to the evolution of many fields, especially Artificial Intelligence (AI)[3]. Since the industrial revolution [163], science and mechanization have become central to many academic challenges, forcing a paradigm shift away from philosophy to focus on systems engineering techniques[4].

The CI domain has evolved over the past 60 years [320] with many new fields of study emerging to dissolve the stiction of obstacles encountered. Many of these relate to attempts at personifying attributes of human behaviour or the knowledge processed within an agent system. An example is the introduction of Multi-Agent System (MAS) with a Beliefs, Desires, Intentions (BDI) framework which form the core to solutions of many real-world problems. This research also suffered impediments, but after a number of recursive attempts, science prevailed by combining several existing techniques or using new technology. For example, complex adaptive agents are required to automate many processes, especially those involving human control. A layered framework capable of providing communications, negotiation and trust are also required to provide the autonomous functionality required to solve the incompatible nature of today's network centric information flows. The incorporation of learning, coordination and cooperation is also required to enable

---

[2] Techniques are the procedures used to perform a specific task and mastery of this function is considered a skill.

[3] The fundamental comparison used to measure this connection is a specific point along the space-time continuum.

[4] It could be argued, one is an extension of the other, however fundamentally the emphasis should be on the doing rather than the thinking.

successful human-machine interaction sought by many scientists conducting Intelligent Agent (IA) research within BDI MAS. Work in this field is emerging as the way forward to solve many interaction problems for agents, MAS, Teams and complex systems.

The question of mechanization calls for automated processes. This problem extends the research of others and requires further research and development effort to expand the proposed capabilities [219]. Automated connectivity between agents has been attempted under controlled conditions; however, this focuses in on a single domain that is constrained within a specific context or framework using a series of experiments. In this context, multi-layering and heterogeneity are chosen as desirable principles in many frameworks, including those using communications and learning capability. For instance, it is natural to implement autonomous characters in multi-media as autonomous agents due to the need to personify the entity and achieve autonomous responses [356]. Using a fusion of agents in a team hierarchy, a holistic, autonomous *being* can be generated, based on the real physical constraints of the environment created. To test these concepts a generic structure embodying the TNC functionality was created. This structure must be capable of being dynamically expanded during run-time using agent pools. The pool should support a variety of capabilities assigned to enable agents teams. These teams should interact with other agents, teams or systems distributed anywhere across a network. Trust is a key characteristic used to mediate this interaction, therefore a discussion on building and measuring trust is required.

This research reviews the fields of AI and AIP for their applicability to support the implementation of the proposed framework. AI techniques and AIP technologies are expected to be able to handle the complex interactions; however, agents have been acredited as the mechanism capable of maximising trust, automation and adaptability within MAS teams. In particular, Distributed Problem Solving (DPS) and MAS provide the technology basis for an overall system architecture of agents with adaptable skills that can be dynamically instantiated or extended by providing adaptable capabilities to satisfy a task. Agents may also combine with other agents, as teams of agents, stand-alone systems, or systems of systems to facilitate the completion of more complex goals when required.

## 1.6 Definitions

### 1.6.1 The Software Evolution

With advances in mathematics and mechanical research related concepts, a new domain that promised to revolutionize the field of computing started to emerge. In 1955, McCarthy labelled this domain as the study of *AI* [243] to describe a conference being convened the following year at *Dartmouth*. Although in a recent interview, he explains this term should have been CI [246]. Minsky espoused that CI is the science of making machines do things that would require intelligence if done by man [254]. This evolution was associated with research conducted during World

War Two, where humanity experienced a series of technological *leaps*[5] that arrived like wave or avalanches and those without the tools (means), technology (method) or training (know how) were swept aside.

Many researchers regard CI as more than engineering and demanded the definition expanded to include the study of science about human and animal intelligence. Intelligence is still considered to be the cognitive aspects of human behaviour, such as perceiving, reasoning, planning, learning and communication. The concept of AI at that time was documented by Newell and Simon [274, 269] who highlighted their production systems [362] in those examples. The field soon divided into two streams with John McCarthy and Nil Nillson considered the *Neats*[6], while Marvin Minsky and Roger Schanks where considered the *scrufs*[7].

Russel and Norvig entered the argument by describing the *environment* as something that received input and provided output, using *sensors* and *effectors*. The sensors were used to feed data into a program that calculated outputs that could be *acted* upon by someone or something. The AI community now uses this notion[8] as the basis of definition of an agent [116].

Computers and computer languages have commanded a significant degree of attention from both industry and academia. Over the last two decades, CI [138, 47] has made a great deal of progress in many fields, such as knowledge representation, inference, machine learning, vision and robotics, however, no single language has been able to achieve supremacy industry wide. Currently three major streams exist that generally support: industry (Cobol/Fortran), programmers of Object Oriented Languages (OOL) and beginners (Visual Basic/Delphi). The later generally provides a front-end for Object Oriented libraries and businesses rely on Cobol because of its large code base. Most academics believe that OOL will continue to dominate the software evolution debate. Given this dominance, ongoing advances in communications technology and an exponential number of users adopting the Internet; platform independence, interactivity and bandwidth are becoming the new agenda.

Reconfigurability refers to the ability of agents to reorganize and form a new subgroup or change their position. Factors of reconfigurability include the MAS environment and communication topologies. For example, if the environment is small, it would limit the movement of agents. Depending on the communication topology, certain agents might be able to communicate only with agents that are linked to them. Communication topology is discussed further in chapter 5.

## *1.6.2 Grammar*

The theory of formal languages encompasses the rules, symbols and functionality. They are commonly measured against their conformity based on semantics,

---

[5] These advances appear to be fuelling achievements at an exponential rate, although population growth and living standards may distort this ideology causing sporadic surges in access and adaptation.

[6] Who started using formal logic as a central tool to achieving AI.

[7] Who retained a psychological approach to AI.

[8] Software that creates an environment which reacts to sensing (inputs) and acting (outputs).

syntax and ontology. Kleene Stare used the term $\Sigma*$ closure to declare a complete set of characters embodied within a language. A Context-Free Grammar (CFG) or Context-Free Language (CFL) embodies the concept of push down automata. The conversion of a high-level language into one that is machine readable, is now called a compiler. These are generally expressed in Backas-Nour or Backas-Normal Form (BNF) (from ALGOL)[9]. The use of Trees or a variety of terms[10] can be used to generate the linguistic, compiler and mathematical logic designs. The Lukasiewicz Notation[11] is used to derive an unambiguous tree by reading a tree down each node successively, Left to Right. This function is seen as the second derivative (Parenthesis-Free) using a hierarchy of operators (called infix, inline or prefix). Any word that can be generated by a given CFG by some derivation also has a left most derivation.

### 1.6.3 Program Languages

Most language authors agree that it is difficult to define a programming language as a single entity. Sammet prefers to define languages through their characteristics. He states, that in his opinion, a programming language is a set of characters with rules for combining them which is architecture opaque, machine independent, require compilation/translation, and be problem-oriented [323]. Many languages are capable of achieving this definition; however, they may lack other attributes, such as; simplicity, speed (performance), robustness, portability and security. Unfortunately, it is not until a language has been in use for a while that its advantages/deficiencies are ascertained [323]. This chapter introduces the modern problems associated with Object Oriented languages and the Internet. Some major problems include; standards, access, bandwidth, interactivity and the ability to extract (pull) the relevant data from the Web. It is not possible to examine all these subjects in full, therefore this chapter will briefly discuss the origins and relevance of a language such as Java, its effect on the Internet and bandwidth.

Researchers have created numerous scripts, macros and entire programming languages as tools to make the process of achieving their task more efficient or productive. Most fail to endure and eventually fade when the milestone is complete. With each new generation of computers, the capacity of existing languages expands due to previously unsolved theories or paradigms. Current programming technologies expose a drift towards component architecture, cross platform scripting languages and in-built *intelli-sense* environments, especially in a networked environment. This shift is primarily caused by *modern users* who will no longer tolerate non-reactive linear applications. Historically there has not been a single operating system, programming language or platform capable of satisfying this demand. Simply examine the evolutionary changes in how we program since the introduction of the assembly language. Programmers stopped decoding numbers and instead thought about

---

[9] Named after John W Backus but published by Peter Naur.

[10] These are also called Syntax Trees, Parse Trees, Generation Trees, Production Trees and Derivation Trees.

[11] A Polish Logistician.

words [92]. As time passed, assembly languages proved cumbersome and required very disciplined development teams to generate even small applications. Programmers readily migrated to higher level languages such as Cobol, Fortran and Algol. This programming evolution led to many subsets and dialects of new or existing programming languages.

Since the mid 1950's, we have averaged one new languages every year, although not all have endured. Many have contributed to the advancement of new languages and eventually a chain or family of languages resulted [154]. Algol influenced the development of many other languages, such as CPL, Simula and Pascal. CPL became BCPL, and Algol was eventually transformed into Pascal. Algol also influenced Fortran, Ada and Modula-2. At the same time, Thompson evolved B from BCPL [365], and in 1972 Ritchie extended 'B' (C without types) into 'C' [191]. A decade later (1983), Stroustrup [352] introduced 'C with classes' which is now known as C++. The latest derivation in this line, is Java, which is based on the syntax of C++ and modelled after SmallTalk [185], UCSD Pascal [4] and Objective-C [66]. Initially C was developed as a low-level tool for systems programmers; however, due to its limitations, higher-level languages based on new paradigms soon developed. As projects grew, many useful 'Structured Development' techniques also emerged [92]. As the number of languages grew we began classifying them as either 'imperative' or 'declarative'; although, more recently new categories have been introduced. Given that the taxonomy of Java is divided between Imperative, Concurrency, Object-Oriented, and Visual categories it is classified as a network-centric Object Oriented paradigm. Table 1.1 shows an incomplete listing of those languages (in no specific order).

**Table 1.1** Languages that Influenced Java

| Assembler | Cobol | Fortran | Ada |
|-----------|--------|--------------|-------------|
| Algol | BCPL | B | C |
| Pascal | Prolog | Visual Basic | C++ |
| Logo | Forth | Lisp | Objective-C |

### 1.6.4  Programming Styles

Formal programming languages used a traditional serial approach, that when written properly involved a top-down style of development. Algol, Cobol, Fortran, C, Pascal and Basic were examples of this McCarthyism approach to logical problem solving. More recently Lisp, IPL, Prolog, Logo, Small-talk, C++ and now Java all contributed to bottom-up approach, or McCarthyism design to programming. Hoare [155] and Wirth respectively, published a set of acceptable guidelines surrounding the characteristics of programming languages [279]. A series of follow-on articles that discussed the merits of categorization by function or use. Operational or Logic based approach compared to Denotational or using the Mathematical or prescriptive language, based on numbers, symbols, functions and equations (Frege's - Calculus derivative [118]), while Wilensky choose to align a High-level (functional languages like Lisp) and Low-level (declarative languages like Prolog) approach to programming [3, 398]. Programming comprises of three specific phases; these include constructing a means to an end (*answer* the problem and provide the appropriate *operator assistance* required to achieve their goals), Documenting the software (*clear, concise code* that is *easily interpreted* and/or extended) and contain a Debugging methodology/plan to *reduce confusion* during the application development. Hoare's principle include: simplicity (modularity), security, fast translation, efficient execution and readability. He cites Lisp *like* applications as possessing these qualities [155]. Formal programming techniques have developed as languages mature. Top-down programming was considered as being the purists (and only) approach to programming, however with the introduction of type safe languages and Object Oriented Programming Languages (OOPLs), bottom-up programming has become a dominant approach to software engineering, especially software component development. This was demonstrated by Zelle et al. [421] using Inductive Logic Programming research into *recursive learning problems*. Here the program flow was written using a top-down approach with recursion used to solve the problem from the bottom-up.

### 1.6.5   Top Down (Formal)

As this topic is covered in many formal programming courses and texts, only a summary is provided to refresh the perspective taken in the book. This strategy involves a sequential task *decomposition* constructing a series of top level *black boxes* which are progressively refined in detail as the software develops. Miller claims that he developed the structured programming concept while working at IBM [402]. A menu system is one example of a top-down application that can be traversed without implementation. As the software develops, empty dialog boxes or confirmatory text messages can be replaced with meaningful results.

### 1.6.6   Bottom Up (OOPS)

Again this topic is covered in many OOPL courses and texts. Thus only a summary is provided to refresh the perspective taken by the authors. This strategy involves a synthesis action that abstracts away the individual problems that are generally composed using the top-down techniques. Each of these solutions can be *assembled* to reveal an emergent application. It became popular in the 1980's as Object Oriented Programming (OOP) diversified the programming community, introduced wide spread adoption of code re-use and pattern designs. Component programming is one type of bottom-up construction. Java Bean *components* are provided in many forms. Most of the simple ones are built from Abstract Windowing Toolkit (AWT), JavaBeans Activation Framework (JAF), Swing and the Bean wrapper, such as Graphical User Interface (GUI) style buttons.

### 1.6.7 Dynamic Functionality

OOPL and component technology are not new. Having matured over the past twenty years, applications have been modified to alter their functionality based on a specific input or results generally determined during run-time. To achieve such flexibility nested IF-THEN statement blocks, Case/Switch statements and using OOPL polymorphism (based on abstract data typing) can be incorporated during design time. Hence a Multiple Document Interface could conceivably be used to open different data types *IF file_ext = txt ......* (insert other extensions that has the appropriate handlers, filters and interfaces). Alternatively type could be set to *document_type = txt* and a polymorphic class would handle that type. This class can be coded as a component and the application assembled to instantiate a pool of objects capable of dynamically responding at run time. The links can be created during design-time or using dynamic allocation at run-time using an underlying messaging system such as Dynamic Composition for Autonomous Object Platforms (DC-AOP). This style of application is commonly found in distributed systems, configured for point-to-point operation using Web Services. Dynamic Composition of Functionality (DCF) services have also been used in mobile agent applications. They exhibit scalability, dynamic functionality and scalable performance with robustness. Kim et al. [194] developed a demonstration to illustrate this process based on enabling research into code and data mobility (based on dynamic loading/binding, reflection and serialization). Examples based on *Messengers, Aglets, Java-to-Go and Voyager* citing the advantages and disadvantages of each approach. Although these techniques may satisfy this theory in a given context, it does not reflect the true focus of the achievable gains. The issue should revolve around how technology (both hardware and software) could be improved to minimize these switching overheads. For example a richer set of registers that enables processes to switch without the need to preserve the current CPU state or the use of a dynamic pool of agents with predefined functionality that can be attached to the present application/process and released for use by subsequent entities.

### 1.6.8 Context Switching

This can be achieved by the temporary allocation of one or more agents (from a pool of agents with specific capability) to a superior, controller or supervisor. Each capability would have its own management function and behave similar to the agent factory design pattern. There would be an initial pool allocation that is capable of growing and shrinking based on the current load. Retaining agents in memory may initially appear a waste of valuable resources, however memory is cheap, especially when comparing the CPU utilization required to spawn, switch and destroy agents or entire processes. This rationale would require changes in silicon, however the concept of accessing agents internally, with the prescribed functionality, using a single CPU, with one or more processing cores on the same motherboard is not available. If this concept incorporates a matrix of CPUs on the same motherboard or distributed among two or more machines would be ideal. Although processor speeds have become relatively static over the past decade, they are being released with multi-core architectures. At present the focus is oriented around creating more efficient applications.

Context is closely related to the extended ability to dynamically switch functionality. Again the problem is not new and significant research has been expended in solving this issue. Work by Doyle on Truth Maintenance System (TMS) [89], McAllester on Logical Truth Maintenance System (LTMS) [242] , Kleer on Assumption-based Truth Maintenance System (ATMS) [82] and Williams and Nayak on Incremental Truth Management System (ITMS) [266] have been used to demonstrate methods capable of improving context switching using software. Williams and Nayak successfully demonstrated a Measures Of Efficiency (MOE) of seven, however they failed to improve the actual switching dynamics, requiring between 100% CPU utilisation (it should be acknowledged that this figure dropped from 700%).

All software comprises of a unnecessarily complex combination of corporate logic and scenario data that forces additional processes to needlessly disrupt the program flow. This hard coded context data is slowly being segregated. Minksy [254] first indicated the concept of separating the cognitive functionality from the deductive or propositional logic. Event management can be optimised and dynamically adjusted using external data relating to context and scenario. The functionality can rely on dynamic components, enabling re-use, robustness, improved maintainability/extensibility and a common reference point (that can assist in measuring outcomes). This problem extends to all entities employed in processing any task. Many applications are monolithic, hard code or avoid the separation of context completely. To solve real-world problems, context is essential and being able to switch between environments enables code re-use and CPU utilisation that assist in rationalising an applications complexity.

## 1.6.9  The Java Language

Java is a new network centric programming language marketed to provide dynamic interaction in a heterogeneous, distributed environment such as the Internet [135]. Like most modern languages, it is supplied with a predefined set of libraries that are supported by the developers to ensure the uniform implementation of the language across all platforms. At present Sun has provided support for Unix, Macintosh and Windows NT/95 platforms; however, many commercial interests are engaged in projects to port the *Virtual Machine* (interpreter) to other platforms. Sun has also enlisted the support of Apple, IBM and Microsoft, who are expected to embed the Java Virtual Machine (JVM) into their current and future Operating Systems (OS). Historically Singleton [340] reported that James Gosling started the development of Java in 1991 [135]. Java originated as a project called *Green*, which was implemented to create a heterogeneous network of electronic consumer products. Attempts to focus on the Personal Digital Assistant (PDA) market saw this project retitled to *Oak*; however, this eventually drifted to become Java. This variant was developed to demonstrate the concept of a *set-top-box* that controlled cable TV or accessed the Internet [221]. Java therefore became a small, reliable, portable, distributed, real-time operating environment. Its syntax and functionality were influenced by techniques presented in existing languages that included C++, Eiffel, SmallTalk, Objective C, and Cedar/Mesa.

The use of any language has led to subsets and dialects. As such, we have seen the expansion of a number of languages into the maturation of new families [154]. The latest derivation of this evolutionary chain is Java, which is based on the syntax of C++ and modelled after SmallTalk, UCSD Pascal and Objective-C.

Object Oriented languages have painstakingly evolved over the last thirty years and this has resulted in the re-definition of this evolutionary metaphor with each new language introduced. Eiffel, SmallTalk and C++ are current examples of object oriented languages that have endured this evolution; however, one or all of these languages will surrender dominance or be superseded by Java. As with the above examples, Java supports a message-based paradigm of inter object communications, but does not support Multiple Inheritance. Proponents of each idiom support diverse views on individual features, especially the implementation of inheritance or the adoption of multiple inheritance. Inheritance is a *revolutionary* mechanism that enables programmers to design code that is reusable and extensible. Java was designed to produce a language that can be deployed into a heterogeneous networked environment and be compiled as an architecture neutral format [135]. The Java team decided not to include multiple inheritance and chose to implement *interfaces* (similar to IDL) to achieve a similar functionality because it is considered easier to master. This is a powerful feature and Java uses *interfaces* to separate *design* inheritance from *implementation* inheritance [221]. Many proponents agree on the importance of inheritance and very few refute that the behaviour of multiple inheritance more accurately mirrors real-world system problems. Others, such as [298], attempt to devalue the object oriented paradigm. He claims that it is still an immature technology that focuses on component reuse, which is no different to what you can do in COBOL or with C subroutines. This may have been true twenty years ago; however, as technology enables programmers to employ new paradigms, this theory becomes redundant. There are many schools of thought about programming languages, their relevance and ability to outperform challenges. This book reveals how current environmental features led to the development of the Java programming language and how they will effect its future.

The teaming concept is generally achieved using multi-layered groups of agents, each with a unique scope, prescribed functionality or responsibility. Multi-layering and heterogeneity are desirable principles of autonomous embedded agents. FreeWill+ is an example of a structured agent used to produce animated graphics that communicate and present the ability to learn [356]. Like all *animated* creatures, these agents need to reflect the capacity of possessing an *artificial mind*. Given the difficulty of this task, the *FreeWill+* designers chose to embed intelligence based on

augmented levels of automation. Such features could include environmental awareness, purposefulness, attitude and the illusion of life. Using a fusion of agents within
a team hierarchy, a holistic, autonomous entity can be generated. This entity could
be based on the real physical constraints of the environment being created. When
overriding supervision, it stifles a level of individual autonomy at the expense of the
team and its goals (such as flocking or crowd dynamics).

*"Societies need rules that make no sense for individuals. For example, it makes no difference whether a one car drives on the left or on the right. It only makes sense when there are many cars on the road [254]!"*

Marvin Minsky

# 2

# Innovation in Modern Artificial Intelligence

This chapter defines the related terms for this topic and key issues surrounding the evolution of the science supporting the CI domain, together with an introduction of several of the tools and training practices developed to support the research in this area. World war two introduced many new technologies and expanded the engineering domain so rapidly that any impediments in a specific research topics were being abandoned in favor of more productive exploits. Over half a century has passed and we are still using the same fundamental computing architectures. Many disciplines have contributed to the development of agents, threads and component architecture. This chapter briefly discusses many of the key developments in CI and their relationship to this book.

## 2.1 Background of Artificial Intelligence

Over the last six decades, AI [47, 138] has made a great deal of progress in many sub-areas such as knowledge representation and inference, machine learning, vision and robotics. The capability of intelligent agents to autonomously perform simple tasks has aroused much interest. The key characteristics that make agents attractive include their:

- Ability to act autonomously;
- High-level representation of behavior - a level of abstraction above object-oriented constructs;
- Flexibility, combining pro-active and reactive behavioral characteristics;

J.W. Tweedale & L.C. Jain: Embedded Automation in Human-Agent Environment, ALO 10, pp. 15–31.
springerlink.com                                                      © Springer-Verlag Berlin Heidelberg 2011

- Real-time performance; Suitability for distributed applications; and
- Ability to work co-operatively in teams.

Patrick Henry Winston first developed a model concerning the representation of 'objects in a blocks world (modular)' which was subsequently called *frame theory*. General problem solving programs followed an ends-means approach/analysis. Backtracking from nodes/lefts along branches back along a tree towards the root after reaching the solution.

**Table 2.1** Brief History of Computational Intelligence

| Year | Advancement | Pioneer | Comment |
|---|---|---|---|
| 500 BC | Abacus | Soroban | China and Japan |
| 700 BC | Roman Numerals | Ancient Rome | Adapted from Etruscan numerals |
| 1614 | John Napier | Logarithms Tables | |
| 1620 | Edmund Gunter | Slide Rule | |
| 1620 | Slide Rule (SR) | Edmund Gunter | |
| 1621 | Two Foot List SR | William Oughtred | Mechanical Adding |
| – | Sliding Scale (SR) | Amedde Mamheim | By Algining Offset Scales |
| 1943 | W. McCullck and W. Pitts | Neuron Emulation | Concept Released |
| 1947 | A. Samuel | Game Play | Checkers (branch & prune) |
| 1949 | Weavcr | Introduced Interlingua | |
| 1950 | J. C. Shaw | AI Programming | Logic Theorist |
| 1950 | Warren | English to French | Software Translator |
| 1950 | A. Bernstien | Chess | |
| 1952 | A. G. Oettinger | Russian to English | Look-up tables used |
| 1956 | R.H. Richens | Semantic Net | Procedural Representation |
| 1957 | H. Simon | General problem Solving | With Constraints |
| 1964 | J. Lederberg | DENDRAL | Chemical Molecular ES |
| 1970 | J, Slagle | Expert Systems | SAINT and MYCIN |
| 1971 | Wilks | Wilks Template | Frame Theory |
| 1976 | D. Slate and L. Atkin | Chess 4.5 | Program beats chess master |
| 1979 | Simon Newell | Information Processing | SOAR |
| 1980 | R. Duda and P. Hart | PROSPECTOR | |
| 1982 | J. D. Foley and A. VanDam | Pixel Vision | |
| 1982 | H. Pople and J. Myers | INTERNIDT | Brute Force Data Analysis |
| 1983 | $5^{th}$ Gen Computers | KIPS | Knowledge processing |
| 1984 | E.A. Feigenbaum et al. | MIS | Data, strategy and planning |
| 1984 | B.G. Buchanan | MYCIN | Expert System |
| 1989 | Klein | RPD | Scripted Logic |
| 1997 | Tambe | STEAM | End means |
| 1997 | Lucas et al. | JACK | DBI |
| 1998 | d'Inverno et al. | dMARS | Reasoning |
| 2004 | Italiax | JADE | Framework |
| 2010 | UniSA | Web-MAT | Hybrid Architecture |

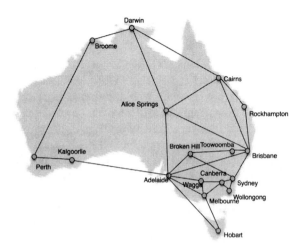

**Fig. 2.1** Problem definition based on map of Australian cities/towns

Simply speaking, an AI application represents a system that interacts with an environment and reasons to generate responses based embedded knowledge. Based on the premise that only one researcher realises one new technique every year, there would be at least 60 examples available to illustrate the developments within AI during this era. Table 2.1 provides a brief list of some of these milestones. A variety of activities may be combined with algorithms that conduct decision making, reasoning, learning, planning, speech recognition, vision, and natural language processing [166]. For example, knowledge can be represented as a set of 'IF ... THEN' statements. In this case, the rules consist of a set of conditions and a conclusion. If all of conditions are true, then the conclusion holds. When reasoning, an expert system (knowledge-based system) will apply many rules. The conclusion from one rule can form part of the IF condition of another rule.

AI is a science that focuses on the problem, not just the solution. Computers can solve tedious problems, however they are poor at representing most real-world problems. It is not good enough to reproduce human-problem-solving techniques, as many would not choose an iterative (search-based) method and search problems don't clearly demonstrate how the brain would solve a problem.

The best solutions are considered complete when they can solve problems within minimal time and space. Search problems use a hierarchy tree or graph structure starting from a parent node. Graphs can be searched width or depth first, prior to extending in the opposite plane. A map of major Australian towns and cities has been created in Figure 2.1, together with its links to adjacent towns or cities. Therefore the first row shows that direct routes exist between *Adelaide* and *Alice Springs, Brisbane, Broken Hill, Hobart and Wagga Wagga*. The second row shows *Alice Springs* associated to

*Brisbane, Cairns, and Darwin*, and so on. From the map a graph of these associations can be built to represent the problem.

### 2.1.1 Modern Artificial Intelligence

Turing [369], Newell and Simon [274], McCarthy [243], Minsky and Papert [255] however more recently Russel and Norvig [320] present the most comprehensive subject matter across the domain, with Wooldridge et al. [410] has taken the lead on Agent technologies. Using agents, human intervention can be minimised, reducing delays or avoid the need to engage in redundant and/or dangerous tasks. Factories, operating machinery and the battle space are all fields where automating tasks is suitable, for instance; flying aircraft in uncontrolled airspace to conduct surveillance [109]. Automation can reduce the cost of operation and prevent loss of life. For example, agent learning and teaming architectures could be used to control Unmanned Aerial System (UAS) during extended or dangerous missions.

The first recorded conference relating to the science of IA dates back to Dartmouth in 1958. Most of this research involved the progressive use of technology in science to collect and interpret data. The wealth of data became unwieldily, forcing researchers to explore data-mining, warehousing and KBS techniques, however the key research domains remained focused on problem solving using *formal/structured* or *reasoning* systems [25].

The growing *density* of data had an overall effect on the efficiency of these systems. Conversely a series of measures were created to report on the performance of Decision Support System (DSS). Factors such as accuracy, response time and explainability were raised as constraints to be considered before specifying *courses of action* [84]. This era[1] was accompanied with an expansion in research into a variety of intelligent decision support systems that were created to derive greater confidence in the decision being generated [417].

Since the eighties, AI applications have concentrated on problem solving, machine vision, speech, natural language processing/translation, common-sense reasoning and robot control [313]. The Windows/Mouse interface still dominates as the dominant HCI, although it is acknowledged as being impractical for use with many mainstream AI applications due to operator delays.

An acceptable definition of intelligence involves human like-behaviour and rational thinking, NOT pure perfection or using the Turing test [256][2]. The dominant driver for AI development is for use in commercial application to help humans become more productive, which may simply mean doing *more work in less time*, like performance monitoring using Simple Network Management Protocol (SNMP).

Newell [270, 271, 272, 367] hypothesized that **only** physical symbols could be manipulated to represent intelligent behavior. In this case *Symbols* are tokens used

---

[1] Subsequently labelled the *Information Age* by Jones, 1990 [182, first printed in 1982].

[2] Bigus believes intelligence refers to the agent to capture and apply application domain specific knowledge and processing to solve problems. This may include using sophisticated, complex AI-based methods such as inferencing and learning [25].

to represent real-world objects or ideas, although Russel and Norvig [320] describe symbolic representation in pattern recognition, reasoning, learning and planning systems. Most use forward and backward chaining to deduce outcomes from source data (normally using a form of semantic net increasingly being termed *connectionism*[3]). Even Turing introduced a neural network with connections in his *B-type O-machine* [369, 370].

Applications with embedded automation capabilities using modern artificial intelligence techniques are becoming so efficient that previous human input/control is becoming the bottleneck. The battle space is one field where automating tasks are deemed to be very important [109][4]. An autonomous, dynamic agent system can be used to minimize system complexity and Human Machine Interface (HMI) disruptions.

## 2.1.2   The Problem Space

To achieve real AI, many challenges exist for researchers to conquer. Technology will eventually support existing research and enable many of these to merge. Achieving a machine with a conscious or one that behaves as a *sentient* being during the next decade is unlikely. The tools are becoming more sophisticated, however many researchers have concentrated on game theory. This field initially used searches to solve puzzles or provide solutions. In *Chess* or *Checkers*, the problem space includes the board, all pieces, every rule and the constraints imposed to direct each the players move. As Samuels specifies, each board state is called a *ply* [324]. The Traveling Salesman Problem (TSP) is an example of tree traversal within a problem space. Each move would start from a node and traverse a tree of branches. The solution is obtained when the goal state is achieved using the lowest possible cost. The most common searches developed include [25]:

Brute-force Search: Blind searches that examine every node of the state space for a solution.

Heuristic Search: Informed or directed searches based on variables bound to the game state to improve the efficiency of the search.

Breadth-first Search: A complete, however wastes time and space to achieve a solution.

Depth-first Search: Has a low memory requirement, but may stall and is not optimal [320].

Iterated-deepening Search: Uses the best features of breadth and depth searching. It is complete and optimal, with a minimal memory requirement.

Heuristic Search: A informed search, like hill-climb, A*, constraint and end-means to produce optimal results with comparable memory requirements to depth-first searches [273, 275, 313].

---

[3] Humans have true biological neural networks in their brains, which are connected by adaptive synapses acting as switches between those neurons in a massively parallel clusters.

[4] Such as flying aircraft for surveillance purposes.

Genetic Search:    Relies on a representational problem which uses biological
metaphors that mutate and rely on fitness functions to produce new, stronger population to estimate a result based on
its very large population.

Other topics include: Local search skeleton using noise injection into tabu-lists to
prevent short cycles [20]. Backtracking is the most commonly used heuristic to increase search efficiency [122]. Prolog and Predicate Logic Directed Acyclic Graphs
(breadth-first search) [284, 349]. Each was developed in an attempt to bypass a constraint being presented within the problem space.

### 2.1.3   Software Agents

There are many definitions of an agent. The major reason for this variance is due
to the exponential growth of diversity and functionality. Wooldridges' definition of
weak and strong Agency [410] currently dominates most literature. The weaker notion defines the term *agent* as having ability to provide simple *autonomy, sociability,
reactivity or pro-activeness*, while the stronger notion is more descriptive and agent
refers to computer systems that extend the above properties, as either abstract or personified concepts [50, 126]. It is quite common in AI to characterize an agent using
cognitive notions, such as *knowledge, belief, intention, obligation* [34] and possibly *emotion* [17]. An agent can be seen as either software or hardware components
within a system capable of accomplishing the tasks on behalf of its source [280].

Agent Taxonomy is formally viewed on a number of planes. These include intelligence[5], mobility and autonomy[6]. Some of the processing strategies include: *reactive*
(reflex [36]), *deliberative* (goal directed [265]) and *collaborative* (BDI [34]) frameworks. Lange suggested the following as the most desirable attributes [213]:

- dynamically adapt to load changes,
- encapsulate protocols,
- fault-tolerant,
- heterogeneous,
- operate autonomously,
- overcome network latency, and
- reduce network load.

They can be equipped with other capabilities such as learning, reasoning and mobility, however complexity grows with overheads and sophistication [56, 175]. System overheads could be minimized given the ability to context switch functionality
and communication models using self discovery. For example, a MAS framework[7]
using an AI mixed with supervisors and functional agents that share the process of

---

[5] More formerly termed agency.

[6] This includes the level of functionality performed independent of human supervision.

[7] Where the MAS is a group of agents or human and agents that are interacting with each
other in order to achieve mutual goals [90, 288].

discovery and goal completion[8] [90]. Agent collaboration and cooperation provide the ability of agents to work together in order to solve problems and achieve common goals and can voluntarily cooperative to problem solving. Autonomous agents have the ability to decide when they would interact with other agents if they have positive motivations [406]. This is the reason agents need to collaborate and negotiate with each other within the team to plan and take actions to solve problems based on the knowledge and context (which is segregated in databases) [375].

When agents are forced to interact, it is easy to mis-interpret the shared knowledge provided external of the receiving agent or team. That means, they need a reasoning ability to enable them to make decisions dynamically. Mobility should also be provided to enable agents to move from one machine to another, across networks and perform their duty on the destination machine and return to the user's machine after the heavy duty processing is complete [175]. The advantages of mobility are the reduction of network traffic and communication cost.

As discussed earlier, no single definition provides a suitable description for all agents. The definition of an agent has been influenced by researchers from different prospectives. The following properties have been discussed amongst a number of subject matter experts:

- Russel and Norvig [320] define an agent as an entity that can be viewed as perceiving its environment through sensors and acting upon its environment through effectors.
- Sodabot [344] views software agents as programs that engage in dialogs and negotiate and coordinate the transfer of information.
- Wooldridge and Jennings [408] state that an agent is a hardware and/or software-based computer system which display autonomy, social adeptness, reactivity, and proactivity.

## 2.2 Agent-Based Computing

A project can only be successful where its members have access to the appropriate tools, technologies and environment required to take the research question from concept to delivery. Here the tools are software, the technologies relate to the platforms and architectures being used and the environment encompasses all input and outputs required to facilitate the task to be solved.

Modern AI has been evolving with the introduction of computers during World War II. In 1943 McCulloch and Pitts [247] reported a Boolean circuit designed to mimic a limited number of functions performed by the brain, where Minsky investigated a theoretical approach which divided the domain into procedural (reasoning) and functional (logical) streams [254]. After the introduction of the 'Turing Test'

---

[8] MAS re-configurability refers to the ability of teams of agents to dynamically organize and form a new subgroup configured to map decomposed tasks to achieve the team goal. Factors of reconfigurability include the MAS environment (bandwidth, capabilities, sensors and processing abilities) and communication topologies.

[256] and a primitive definition of AI [246], a series of developments investigated software based deductive reasoning, natural language interfaces and expert systems. As technology matured, Distributed Intelligent Systems (DIS) and the emergence of object based programming facilitated advances in IA [320]. To the best of our knowledge, there has been no comparison between software performance and hardware capacity. During the mid 90's we attempted to ascertain the best measure to benchmark hardware[9]. Using a database to follow price and performance[10], we tracked increased peripheral and data-bus bandwidth, with faster microprocessor and memory speeds. However, regardless of these measures, it became abundantly clear that the user ultimately determines the acceptance of any software or application. The HCI created the most annoying bottlenecks experienced by users, although many of these issues have been solved using advances in technology [277]. Natural and real-time processing still present a number of challenges, especially the environment in which the software is required to operate.

Many people believe that OOPL developed from 'C' when 'Stroustrup' introduced 'C with Classes' [352], however OOPL originated with Simula [72], LISt Processing (LISP) [123, 187] and Smalltalk [185]. The origin is less important than the ability to provide *high level abstraction* in complex systems. Unlike agents, classes are passive in nature. They are simply invoked with state and a functional interface. Classes have no choice about how they operate or when functions are called/executed. Agents have attracted a variety of definitions. In fact researchers have been using a descriptive approach to distinguish their study. Some of these include [228]:

- autonomous [99],
- avatars [357, 113],
- information agents [393],
- Intelligent Agent [409],
- interface agents [240],
- mobile agent [213],
- multi-agents [375],
- software agent [126], and
- virtual agents [79].

By definition, agents have *autonomy* and therefore make decisions on whether to participate in a calculation or requested operation. To achieve *mobility* they are more capable when instantiated within a single thread. To interact with the external environment, agents require *sensors* and *effectors* for both **stimuli** and **control**. Therefore creating agents (based on the concept of classes), requires a divergent software skills discipline, hence a host of Agent Oriented Programming Languages (AOPLs) were

---

[9] Moore's law certainly mapped the growth of capability in the terms of silicon [257], similarly Metcalfe [130] predicted the exponential growth of networks and the resulting explosion of internet connections while a variety of intelligent decision support systems being generated [418] in what Jones labeled the Information Age' [182].

[10] Based on the average configuration for a leading edge computer from the same chain store, costing approximately $3,000.

spawned[11]. Agents have been described with *weak* and *strong* notions [410], although they should technically be categorized as having; autonomy, reactive, proactive and social abilities [281]. New dimensions to be considered could include: taxonomy [31], semantic [147, 241] architectures, although creating a new category for the sake of labelling research should be minimized. The same concept should be adopted for architectures and frameworks.

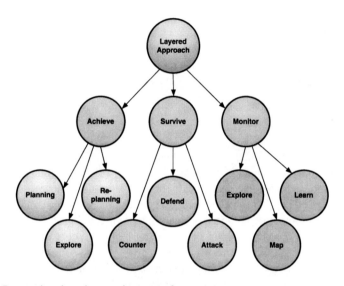

**Fig. 2.2** Remapping the sub-sumption example

## 2.3  Agent Categories and Architectures

Wooldridge [412] believes that there should be four categories of agent; reactive, deliberative, hybrid and distributed agent architectures. Brooks [36] on the other hand developed what he called a subsumption architecture while Maes [233, former colleague of Brooks] postulated that intelligent behavior as being an emergent phenomenon. The subsumption architecture is an extension of reactive agent systems while the Agent Network Architecture (ANA) combines a 'society of mindless interacting systems' which are essentially MAS [228]. The definition in each case should have been expanded to include the new research, in lieu of an attempt to single out these enhancements. The subsumption model uses a serial processing concept in lieu of an object-oriented approach. It uses a logical 'case' style network to switch behavior. Brooks focused on a primary capability of *explore*, decomposed into a number of tasks that are required to achieve a goal. When the same problem is represented graphically and the implementation created using an object-orient approach, bidirectional chaining could be achieved providing greater efficiency (see Figure 2.2).

[11] A non-exclusive list of examples examined include: Collaborative Agent for Simulating Teamwork (CAST) [418], JACK [161], MadKit [106, 361], CIAgent [25], Procedural Reasoning System (PRS) and distributed Multi-Agent Reasoning System (dMARS) [87].

By expanding the Brooks subsumption model, and applying an AOPL (based on the OOPL architecture), the *reuse*, *scale* and *robustness* objectives are visible. Each agent becomes a capability [catalogued within a list] is held in a container[12], which *adapts* its response (effectors) in accordance with its stimuli (sensors) while operating *autonomously* in its environment. Many agree with Nwana [280] who uses four categories, but we believe these could be reduced to three, by combining the reactive and proactive categories as adaptability. Most researchers try to distinguish their work by using a new label or title. Bratman [34] for instance used BDI [128, 303], Boyd [144] Observe Orient Decide and Act (OODA) [64] with many researchers [98] using learning and re-planning [14] which essentially all stimulate decision making. This means that a challenging new researchers could conceivably transform all these terms into a Stimulate Observe Decide and Act (SODA) loop.

### 2.3.1   Multi-agent Systems

Research on MAS is very active with applications being developed to assist in many human-computer environments. MAS inherit many Distributed Artificial Intelligence (DAI) motivations, goals and potential benefits, and also inherit those of AI technologies. As described in [280], agent researchers concentrate mainly on deliberative-type agents with symbolic internal models, dealing with interaction and communication between agents, the decomposition and distribution of tasks, coordination and cooperation, conflict resolution via negotiation and so on.

A multi-agent system may be regarded as a group of ideal, rational, real-time agents interacting with one another to collectively achieve their goals. Each agent needs to be equipped to reason over events within the environment and the behaviour of other agents. Based on this inference, agents need to generate a sequence of actions and execute them.

MAS are mostly targeted to be an ideal solution for problems in heterogenous environments. One of the classic examples of heterogeneous environment is a game of soccer. In this type of a heterogeneous environment, each agent needs to have a strategy to solve a particular problem(s). In this context, a strategy refers to a decision-making mechanism that provides a long term consideration for selecting actions to achieve specific goals. Each strategy differs in the way it tackles the solution space of the problem. The presence of multiple agents necessitates the need for a different treatment. In order to achieve a solution to these strategies a coordination mechanism is required to handle the interactions between agents. This mechanism is responsible for the implementation of the agents' actions, but also the planning process that goes into the implementation.

JACK is a Java based BDI agent architecture that is mature enough to provide decision support for complex reactive systems [195]. The architecture needs some refinement, but with further development, it is suitable for systems where human

---

[12] Which is a thread with basic behavior, however it has dynamic access to functional capability, agile communication protocols or transportable mechanism.

involvement is the more of a critical factor. Examples of such systems have been attempted to generate autonomous adaptive systems in complex reactive environments, such as surveillance and control operators that build situation awareness. To improve this scenario, an MAS capability of coordination and learning could conceivably be incorporated to automate or assist operators in conducting these roles.

In some MAS, agents would be able to send messages directly to another agent in the team where it can be uniquely identified [90]. Other topologies include agents only communicating with agents that are linked directly to them. For example, through a graph network or a hierarchy structure. In a hierarchical structure, there might be one or more *controller* agent that coordinate the tasks of other agents within their hierarchy. One that restricts agent communication for guidance or communication with their supervisor agent. There are advantages and disadvantages of this type of structure. One advantage is that it reduces and simplifies the interactions between the agents because agents only interact with certain agents. However, the major disadvantage is that the communication among other agents in this structure is prone to the failure for one or more agents in the chain, especially as the linkage to a group of agents could be broken, isolating it from the rest of the team.

A MAS can include a group of agents or a group comprising of 'humans and agents' that interact with each other in order to achieve mutual goals [288]. In such a system, it is assumed that the agents may not have full knowledge about the environment of other agents. Therefore, the interaction among agents is an important feature that enables them to use the information provided by other agents and learn more about the environment. However, in some scenarios, no interactions is required among the agents, therefore tasks can be separated and completed independently by an agent [288]. An example of this is an environment where agents are required to find specific targets. Each agent can independently search the environment to find a target based on the criteria given about the targets characteristics. This can be inefficient if agents find an alternative target (currently assigned elsewhere). For this reason context related information must be available to all agents doing that task to enable agents to interact. The problem becomes more complex or inefficient if agents merely interact to notify one another about static information within the environment.

According to Dudek et al. (1993), using MAS has advantages over a single agent. For example, replacing a single complex robot with a group of simpler robots to explore an unknown area. This reduces the design complexity of each robot, is more economical and scalable. The overall rate of failure would also decrease because if one or more robots fail, the whole system would still be able to function.

The shared knowledge among agents should not only contain the knowledge of meaningful words and phrases, but also the knowledge of particular issues or items of interest to the agents [209]. For example, in a MAS, such as UAV swarm, the individual UAV should have a common language among themselves to communicate and also a shared knowledge-base of certain characteristics such as velocity, position, targets and hazard.

When a group of agents interact with each other or other entities they form a MAS [407]. The interaction of agents enables them to use other agents knowledge to learn more about their environment and use other agents intelligence and skills to perform tasks and solve problems. In a MAS, agents can search for other agents and decide to interact and collaborate with them when they have positive motivations in order to achieve common goals [411]. This collaborative problem solving requires agents to recognize the need for cooperative actions. The need for cooperation is recognized by an agent when it realizes that its goal cannot be achieved without another agents resource [375]. The agent would then search for other agents that can assist in accomplishing the goal. This leads to the formation of a group of agents that have the required abilities to solve the problem and they collaborate and negotiate with each other to plan and take actions for solving the problem. Autonomy is the most important property coded into an IA. It provides agents with the ability to make decisions and solve given tasks within the host environment. In many cases it is based on the state or goal that can be achieved without the need of any external system or human intervention [56]. Often this is achieved by repeated trial and error [288]. An autonomous agent needs to be both proactive and reactive. Proactiveness is the agents ability to plan and perform required tasks to achieve its goals [56]. Reactive behaviour refers to the ability of the agent to notice and respond to environmental changes. By having the above characteristics the agents can then socially interact with each other or collaborate to perform a specified task. Social behaviour also enables agents to coordinate, cooperate, negotiate or even compete to achieve an objective [103].

According to Huhns et al. [164], the environment, agents, and inter-agent interactions are some factors for characterizing a MAS. Dudek, G. et al. [90] further discusses characteristics to include swarm size, reconfigurability, composition, and communication topology. When MAS are applied to real world applications, other factors are also considered such as communication range, bandwidth, and the agents individual capabilities such as sensing and processing abilities.

The technology of MAS is becoming more popular these days especially in the field of e-commerce where large-scale open systems are constructed with multiple information sources [405]. In such situation there might be different agents built by different organizations interacting with each other. Therefore, the inter-operability becomes an important issue and the need for having a common language with universally agreed semantic arises. Agents have been defined in a variety of ways as their functionalities have grown in different fields [375].

### 2.3.2 Agent Teams

An agent team can be formed using a variety of architectures, hierarchies and communications models. A single agent in a simulation could be used to conduct a specific function or more complex tasks that may control an entity. Multiple agents can be instantiated as a team of agents that do the same task and share the load on a single platform or even distributed across a number of machines or networks. A team of agents may contain a supervisor that allocate one or more tasks to one or more sub-agents in order to complete a goal as part of the overall teams effort. As mentioned before, MAS consist of several agents that interact with each other and their environment. In a MAS, agents may be connected, but not necessarily programmed to intelligently cooperate with each other. The aim of this interaction is to distribute knowledge and intelligence of agents [90]. An example would be a team of agents processing one or more entities to establish/maintain situation awareness. Different systems may be instantiated with a hierarchy, with each level performing predetermined tasks in subordinate, supervisory or sibling roles. Further research on these concepts have been documented by the authors in articles surrounding of the TNC model [301, 376, 377].

The major benefits of intelligent agent technology is gained during its deployment in complex distributed applications such as virtual enterprise management and the management of sensor networks. However while the agent paradigm offers the promise of providing a better framework for conceptualising and implementing these types of systems, there is a need to recognise the underlying programming paradigm and support specified standards, design methodologies and reference architectures before these applications can be developed effectively. As noted above, these are beginning to appear, but more experience needs to be gained with them and the software community needs to be educated in their use. Given the nature of these applications, a killer application seems unlikely at this level. Rather we would expect to see a gradual shift from the object-oriented to the agent paradigm in intelligent domains as the supporting framework matures.

The underlying theories of cognition will continue to prove adequate for large-scale software developments. The key theories (BDI and production systems) date from the 1980's and have a long pedigree in terms of their use in commercial-strength applications. This longevity indicates that their basic foundation is both sound and extensible, which is clearly illustrated in the implementations; from PRS [114] to dMARS [85], to STEAM [360], to CAST [418], to JACK and JACK Teams [174]. New cognitive concepts may gain favor (In the form of norms, obligations, or perhaps commitment), but we believe that these concepts will not require the development of fundamentally new theories.

While we believe that the existing theories are sufficiently flexible to accommodate new cognitive concepts, we perceive a need to develop alternative reasoning models. In the case of the JACK implementation of BDI, a team reasoning model is already commercially available in addition to the original agent reasoning model. On the other end of the spectrum, a low-level cognitive reasoning model (COJACK) has been recently developed. This model enables the memory accesses that are made by a JACK agent to be influenced in a cognitively realistic manner by external behaviour moderators such as caffeine or fatigue. Interestingly, COJACK utilises an ACT-R like theory of cognition, which in turn is implemented using JACK's agent reasoning model. From a software engineering point of view, it should be the reasoning model that one employs that shapes an application, not the underlying cognitive theory. Thus there is an opportunity through the provision of *higher level* reasoning models like OODA and their incorporation into design methodologies to significantly impact on productivity and hence market penetration.

Agent teaming gained popularity in recent years and was categorised into the prominent domain of MAS. It is believed that three important aspects of the design should include; *Communication, Coordination and Cooperation* because they also play an important role in IA and agent teaming. Multi-agent teaming also takes inspiration from human organisational models of team operation, where role playing such as leadership, communicative, cooperative and collaborative skills empower the success of team.

### 2.3.3   Intelligent Agents

The area of agent technology has expanded to include the concept of intelligence. Intelligent agents are described as computational systems that can interact with each other and other entities autonomously that process proactive, reactive, and social behaviours [103]. Rudowsky (2004) initially connected the study of AI and agent systems [319]. AI involves the study of software and equipment that mimic intelligence (Each is presented with some ability to learn or plan) while the study of agents focus on integrating that intelligence into software or equipment. This distinction may seem implicit, although not all problems within AI are necessarily solved using agents.

Shoham (1997) first proposed the concept of Aspect Oriented Programming (AOP) [335]. The key idea of AOP is that agents mental notions are represented as BDI elements. This architecture is well known and has been used widely in the design of many contextual based MAS. They share the motivations of many DAI goals and benefits, while inheriting the technologies used to support AI. As Nwana (1996) described, agent researchers concentrate mainly on deliberative-type agents with symbolic internal models, dealing with interaction and communication between agents, the decomposition and distribution of tasks, coordination and cooperation, conflict resolution via negotiation. Nwana chose to use a problem solving approach [280] using agent teams categorized as; deliberative, reactive or hybrid agents.

With intelligence and interaction, agents act as a bridge between humans and machines. The development of social intelligent agent is not quite achieved due to the lack of characteristics like learning and teaming. It is needed to include a human in the supervisory loop because the dynamic environment has many external and changing constraints [109]. According to Tweedale et al. (2006) [375], new researchers and technologies are focusing on developing human-centric MAS.

### 2.3.4   JACK Intelligent Agents

The JACK IA framework is a development platform for creating practical reasoning agents in the Java language with a BDI reasoning constructs. It allows the designer to use all features of Java as well as a number of specific agent extensions to aid scheduling and communications. Any source code written using JACK extensions is automatically compiled into regular Java code before being executed. Each agent has beliefs about the world, events to respond reactively, goals that it desires to achieve, and plans that define what to do. When an agent is executed, it waits until it is provided with a goal to achieve or receives an event for which it can respond reactively. It then reasons using its own beliefs prior to response. If a response is required it selects an appropriate plan to execute in order to respond. JACK agents are based on the BDI reasoning model and can exhibit: Goal-directed behavior, where the agent focuses on the objective and not the method chosen to achieve it; Context sensitivity, keeping track of which options are applicable at each given moment using beliefs; Validation of approach, ensuring that a chosen course of action is pursued only for as long as applicable; Concurrency, behaviours in the agent are executed in separate, parallel and prioritized threads.

The language is used to develop agent based systems using agent-oriented paradigms, the language is complete with a compiler, a powerful multi-threaded run-time environment and a graphical environment to assist with development. Beliefs have been implemented as relational databases called beliefsets, however developers can also use their own Java-based data structures if needed. Desires are realized through goal events that are posted in order to initiate reasoning. This is an important feature because it causes the agent to exhibit goal-directed behaviour rather than action-directed behaviour, meaning that the agent commits to the desired outcome and not on the method to achieve it. An intention is defined as a plan to which the agent commits to after choosing from a library of pre-written plans. The agent is able to abort a plan at any time depending on its beliefs and also consider alternative plans.

JACK Teams is an extension to the original JACK framework that provides a team-oriented modeling. The JACK Teams *extension* introduces the concept of *Team reasoning*, encapsulated by agent team *behavior and roles* required to define what each agent is required to do within the team. Using this Teams extension, individual agent functionality is maintained. Team oriented programming is provided by the designer who specifies: the functionality a team can perform, its roles and needs in order to form a team role, coordination and knowledge between team members can also be stimulated.

Roles are bound to agents at runtime. This means that it is possible to have different combinations of agent-role relationships. For example, one or more role can be performed by a variety of different agents or simultaneously by one or more agent.

Belief can be propagated between members of a team. Sub-teams inherit beliefs from higher-level teams and conversely, team members synthesize beliefs up-ward. JACK teams was developed to support structured teams, therefore the structure of roles in teams must be defined at compile-time. Consequently, sub-teams can only communicate and share information if it has been previously defined in their team structure.

### 2.3.5 Collaborating Agents for Simulating Teamwork

A number of researchers have integrated the Recognition-Primed Decision (RPD) model into MAS [196, 418] to capture the decision making abilities [199] of domain experts based on the recognition of similarity between the current situation and past experiences. In the first phase (*recognition*), a decision maker develops situation awareness and decides a course of action. In the second phase (*evaluation*) a decision maker evaluates each course of action. In 1989, Klein [197] introduced a model that evolved in an agent environment under teamwork setting into the RPD Agent architecture [101]. Hanratty et al. describes this architecture of using four modules [145]. The communication manager module governs the inter-agent communication and organizes conversations. The expert system module is a rule-based forward chaining system. It contains knowledge related to the other agents and external world. The process manager module is responsible for scheduling and execution of plans. The collaborative RPD module facilitates the collaboration of humans and RPD agents [199].

The developers of RPD-CAST agent have tested their software in a military Command and Control ($C^2$) simulations involving intelligence gathering, logistics and force protection [145]. Under normal time pressure, the human teams made correct decisions about the potential threat, although the team performance suffers due to the lack of information sharing and MAS resulting in incorrect decisions making. This has been illustrated using an Recognition-Primed Collaborative Agent for Simulating Teamwork (R-CAST) agent system to support human-agents when making decisions. This modular or compontised structure can easily be simulated and is suitable for black box design. The CAST architecture [100] is designed to simulate teamwork. It supports proactive information exchange in a dynamic environment[13].

## 2.4 Blackboard Model

A blackboard system may be thought of as a componentised system, where each box could function as a database, series of *pigeon holes* or behave with an unknown black box behaviour that represents a specific aspect of a system or sub-systems engaging a problem. This needs to occur in an environment where experts and modular software subsystems, called knowledge repositories, capable of representing different points of view, strategies, and knowledge formats, required to solve a problem. These problem-solving paradigms may include:

- Bayesian Networks,
- Case-Based Systems,
- Fuzzy Logic Systems,
- Legacy (traditional or formal procedural) software systems,
- Neural Networks, and
- Rule-Based Systems.

The blackboard architecture concept was conceived by researchers in the field of AI a few decades ago. Originally designed for the HEARSAY-II speech-understanding system [38], the blackboard architecture is one of the most widely used architectures in symbolic MAS. The intent of this research was to address issues of information sharing among multiple heterogeneous problem-solving agents.

The blackboard architecture is not very specific and can be seen as a *meta-architecture*, architecture for implementing other architectures. The name, blackboard architecture, was chosen to evoke a metaphor in which a group of experts

---

[13] The Shared Mental Model (SMM) consists of team processes, team structure, shared domain knowledge and information-needs graphs. The Individual Mental Model (IMM) stores mental attitudes held by agents. The information is constantly updated using sensor inputs and messages from agents. The Attention Management (AM) module is responsible for the decision-maker agent's attentions on decision tasks. The Process Management (PM) module ensures that all team members follow their intended plans [418].

gathers around a blackboard to collaboratively solve a complex problem [95]. The blackboard model is based on a division into independent modules that do not communicate any data directly, but interact by sharing data in a common storage as is shown in Figure 2.3. Blackboards form a good way of communication between heterogeneous information sources. The sources do not need to know about each other, the only thing they see is the blackboard.

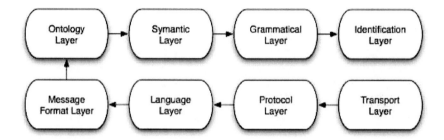

**Fig. 2.3**  Proposed Blackboard Model

*"Although it would appear that we have reached the limits of what is possible to achieve using computer technology, one should be careful with such statements, as they tend to sound pretty silly in five years. [366]."*

John von Neumann

# 3

# Research Directions in Automation

This chapter presents a brief introduction to existing research in Knowledge-Based engineering to achieve automation using IAs. It defines how the term intelligence is determined and the tools used to exploit human machine interaction. So what is Intelligence, why use AI and which architecture provides the best choice of tools. A whole field of science has developed around AI and is based predominantly on computer technology and the enhancements used to develop their capability. The literature illustrates the evolution of the key sciences used to support AI and distinguishes between human, machine and the architectures required to solve problems. Disruptions in developing AI techniques bounded with insertion of new technology, but most were focused on commercial applications, other than niche domains like AI.

## 3.1 A Knowledge Based Approach

During World Wars one and two, technology accelerated to meet the need of cryptographers and code generation. Computers, programming languages and compilers were all optimized to support these efforts. The research community continued to pioneer mathematics and computational problem solving post-war, but industry was already focusing on mass production of automated office applications in the corporate arena.

How can complex adaptive agents be used to enhance communications within dynamic components inside a MAS team to achieve autonomous functionality across distributed intelligent system in todays network centric information space. This concept is illustrated in Figure 3.1. Since the mid 70's agent Teams have been the

J.W. Tweedale & L.C. Jain: Embedded Automation in Human-Agent Environment, ALO 10, pp. 33–48.
springerlink.com                                                    © Springer-Verlag Berlin Heidelberg 2011

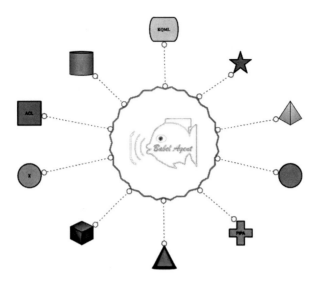

**Fig. 3.1** Concept Babel Fish and Dynamic Interoperabilty

focus of significant research efforts, which still endures today. Techniques have developed in stages with technological enhancement of hardware and software. The future of automation relies heavily on AIPs to provide the essential decision support functions within IA using teams. Due to the human influence, additional techniques must also been considered. Existing experimentation has successfully been used to highlight a number of outstanding bottlenecks and identify avenues of improvement. Trust, negotiation, communications, learning, collaboration, cooperation and coordination including a number of techniques that are available to enhance automation, interoperability and adaptation. The tools have been proven to work in isolated cases or in a limited variety of combinations. On going research, framed around the following assumptions, continues to leverage the existing refinement required to evolve a greater understanding of the topics outlined in the following sections. Much of this research emanates from England, Japan, Europe and America. Other countries are now contributing because of the advances being realized in the *Web Services* field, but little is being done with agent teams, communication or dynamic functionality.

## 3.2  Assumptions

The following assumptions have been made when framing the problem statement:

- It is intended to demonstrate the results of this study using a simulation of how communication streams need to be bifurcated in order to successfully interoperate.
- Not all models used are being designed from first principles.

- The concepts and implementation may be adapted from the existing bodies of research.
- The functionality of the blackboard model can be adjusted to suit the schedule and maintain the study goals.
- The communications methods, languages and protocols will be limited to known standards during the initial prototype design.
- Other assumptions will be added to reduce risk as the literature review and related AIP technologies are assessed prior to implementation.

## 3.3 Knowledge Representation

The Heuristic Computing domain has evolved [320] with many new fields of study emerging as key obstacles are being solved. Many of these are related to attempts at personifying attributes of human behaviour or the knowledge processes into an IA system. During this time, AI [47, 138] has made a great deal of progress in many fields, such as knowledge representation, inference, machine learning, vision and robotics. As discussed, AI is the science of making machines do things that normally require the same intelligence when the same task is done by humans [254]. AI is more than engineering, as it requires the study of science regarding human and animal intelligence. Current AI research embodies the cognitive aspects of human behavior and uses *reasoning, planning, learning* all with an underlying *communications* model. Initially Newell and Simon introduced the *production systems* as an example [362], but, the field quickly divided into two streams. The *Neats* were led by John McCarthy and Nil Nillson, who used formal logic as a central tool for achieving AI, while the *scrufs* led by Marvin Minsky and Roger Schanks, who used a psychological approach to AI. Russel and Norvig entered the argument by describing an *environment* as something that provides input and receives output, using *sensors* as inputs to a program and producing outputs as a result of *acting* on something within that program. The AI community now uses this notion[1] as the basis of definition of an agent [116].

After his football coaching career, Knuth became a mathematician and subsequently a computer scientist. He is acknowledged as being the inventor of the modern computer [201] and has published a significant series of seminal papers based on his wealth of experience in the computing domain. These books document data structures, algorithms and many formalized programming techniques which are still in use today. Wirth formalized the basic requirements of a program, proposing it as it embodies data, data structure(s) and related algorithm [403] (predominantly as separate components; that is *data* and the *corporate logic*). This approach enables the programmer to represent knowledge in a structured form. For each element of knowledge Skills, Rules and Knowledge (SRK) [306], programmers concentrate on First Order Predicate Logic (FOPL) because it can be used to show how anything that exists can be false. It contains Axioms based on *Arity*[2] in predicates surrounded

---

[1] Software that creates an environment that reacts to sensing (inputs) and acting (outputs).

[2] Arity is the number of argument places a predicate or function has.

by one or more universal qualifiers (that can be nested) [96]. Kowalski proved this style of logic (originally concieved by Frege [117]). Herbrand later used this logic to formulate a model based on a domain or a logical view of the world [329]. Horn minimised this logic by negating the model [263] which led to the development of the first prolog compiler in *Edinburgh* during 1977 [189]. The science of AI stalled as the scale of the problem being represented began to encompass real-world problems. Mainstream graph and searching techniques were being employed with limited success. The complexity of representing knowledge, maintained a statistical/mathematical direction. The use of FOPL quickly evolved into *frames, semantic-nets* (briefly exploring uncertainty) and again stalling at *neural-nets* (knowledge engineering and machine learning).

Many agree with Rasmussen's definition of knowledge, "as facts, conditions, expertise, ability or understanding associated with the sensors (sight, taste, touch, smell and sound) relating to anything in their environment [305]". This originally confined the analysis and processing of knowledge as a symbolic representation processes being diagnosed [306, 307, 308, 309]. Early systems were forced to store symbolic tables in flat databases, however the growth in capability of hierarchical, relational databases has extended the scope of knowledge engineering, especially in expert systems [54, 382].

The concept of knowledge is a collection of facts, principles, and related concepts. Knowledge representation is the key to any communication language and a fundamental issue in AI. The way knowledge is represented and expressed has to be meaningful so that the communicating entities can grasp the concept of the knowledge transmitted among them. This requires a good technique to represent knowledge. In computers symbols (numbers and characters) are used to store and manipulate the knowledge. There are different approaches for storing the knowledge because there are different kinds of knowledge such as facts, rules, relationships, and so on. Some popular approaches for storing knowledge in computers include procedural, relational, and hierarchical representations [25]. Procedural representation method encodes knowledge into program code and sequential instructions. However, encoding the knowledge into this form of algorithm, must be preprocessed, which makes the technique difficult to modify. Therefore, declarative knowledge concept is used to represent facts, rules, and relationships by themselves and separate knowledge from the algorithm used to process it. In relational representation method such as Structured Query Language (SQL), data is stored in a set of fields and columns based on the attributes of items. This method of representing knowledge is flexible but it is not as good as hierarchical representation in stating the relationships and shared attributes of different objects or concepts. Network hierarchical database systems are very strong in representing knowledge and *is-a* relationship between related groups. An example of an *is-a* relationship relates to a specific object that is linked by a more abstract term, such as linking the object *apple* to the category of *fruit*. Other forms of knowledge representation used include *Predicate Logic, Frames, Semantic Nets, If-Then rules and Knowledge Interchange Format*. The type of knowledge representation to be used depends on the AI application and the domain that IA is supposed to function [25]. In these cases where there are limited numbers of situations that might occur, knowledge can be hard-coded into procedural program codes. However, as the number of situations increases IAs would need a broader knowledge base and a more flexible interface. Therefore, knowledge should be separated from the procedural algorithms in order to simplify knowledge modification and processing.

## 3.4 Human Factors

Given the body of knowledge surrounding human factors, their design and functional architectures, many common elements influence human interaction with IAs [165, 205, 381]. These include:

| | |
|---|---|
| Workload: | The demonstration and evaluation of the potential benefits of AI for the purpose of enhancing the operational effectiveness and improvement of human-operator workload. |
| Authority: | The characteristic of the support provided to the human-operator is always advising and associative, rather than directive. It is assumed that the human-operator is in charge and remains the final authority for all actions. |
| Tasking: | Functional architectures are all implemented as highly modular structures, where decomposed tasks are automated, based on independent operations. |
| Control: | Dedicated functional modules are allocated to manage and coordinate between the human-operator and machine-assistant. These functions are additions to the non-automated situation and are often called execution manager, information manager, task manager or mission manager. |
| Advanced Information Processing: | The research programs employ a multitude of AIP technologies, methods and tools. The most frequently used are Expert Systems (rule-based systems). Other important technologies are: case-based systems, planning (script-based and plan-goal graphs), fuzzy logic and distributed blackboard systems. |
| Scale: | The typical size of knowledge bases of prototype implementations is substantial. In the case of the Pilots Associate (PA) there are 828 rules for the pilot vehicle interface and 990 rules in total for the other modules. The four management modules of the Copilote Electronique (CE) each roughly consist of 400 rules and 3000 objects [15]. |

These findings and the supporting design philosophy surrounding the Pilot's Associate (PA), CE and Mission Management Aid (MMA) have paved the way for the development of the generic framework for human-machine teaming, especially in an air environment. The fundamentals should be used in conjunction with any architectural framework when developing MAS architectures that require human-machine interaction or oversight.

## 3.5   Decision Support Systems

The concept of DSS emerged in the early 70s and developed over the next decade.
A good example of a DSS is a closed system that uses feedback to control its output.
According to Russell and Norvig, a thermostat could be regarded as an agent that
provides decision support [320]. DSS are computer programs that assist the users in
decision making that incorporate data models which support humans [104]. They
are more commonly employed to emphasize effectiveness. This gain is generally
achieved by degrading the systems efficiency, however using modern computing, this
factor is less of an issue. Russell and Norvig also defined an agent as "anything that
can be viewed as perceiving its environment through sensors and acting upon that
environment through effectors [320]" noting that a DSS generally forms the basis of
components within an agent, application or system.

Agent oriented development can be considered as the successor of object oriented
development when applied in the AI problem domains. Agents embody a software
development paradigm that attempts to merge some of the theories developed in AI
research within computer science. The growing *density* of data had an overall effect on the efficiency of these systems. Conversely a series of measures were created
to report on the performance of DSS. Factors such as; accuracy, response time and
explain-ability were raised as constraints to be considered before specifying *courses
of action* [84].

Since the eighties, AI applications have concentrated on problem solving, machine vision, speech, natural language processing/translation, common-sense reasoning and robot control [313]. The Windows/Mouse is currently the predominant HCI, it
is  acknowledged  as  being  impractical  for  use  with  many  mainstream  AI
applications.

Intelligent data retrieval/management relies heavily on the designer and/or pro-
gramer(s) to provide the sensors, knowledge representation and inference required
to provide a meaningful output to stimulate the operator(s). Scholars believe that
operators respond symbolically using 'Thin slicing' to provide judgement or derive
snap decisions [88][3]. Through his work on decision making under pressure situations, Klein [197, 198, 199, 200, 196] extends the concept of processing information gained through human based sensors against the experiential patterns stored in
our subconscious mind. To maximize this concept, the combination of both issues
(clouded judgement and subconscious expertise), Boyd [64, 144] further extends the
human thought process to enable us to employ a control mechanism to focus on the
goal, through OODA. Being a closed loop system, stimuli can be used in place of observation of OODA in a sensor based OODA system termed SODA, especially in a
known contextual environment. Such issues predominantly surface during complex,
hostile engagements in an environment where BDI can result in *mode* confusion, po-
tentially compromising set goals [33, 85, 303] . To reduce this problem Rasmussen

---

[3] This is often impaired where verbal queues are used to describe the symbolic links that are
established.

proposes that we should use an experience ladders [304, 305, 306, 308] based on the rules, skills, knowledge associated with the context of the environment and the scenarios that can be extrapolated by SME. Vicente studied this approach from the *work domain* perspective, concentrating on the *cognitive domain* to derive the problem scope [388].

## 3.6 AI in Decision Making

The application of AI in decision making is not new. Recent advances in AI techniques provide better accessibility to this technology which has resulted in an increased number of applications using DSS based MAS. These applications aid the decision maker in selecting an appropriate action in real-time, especially when under stressful conditions. The net effect is reduced information overload by enabling up-to-date information to be used in providing a dynamic response. Intelligent agents are used to enable the communications required for collaborative decisions and deal with uncertainty. AI researchers possess a comprehensive toolbox to deal with issues such as, architecture and integration [231]. A number of recent literature are listed in table 3.1.

IA are perhaps the most widely used or applied method in decision making applications in recent years. This utilization has significantly advanced many applications, particularly Web-based systems (see for example [293]). Learning can be incorporated into agent characteristics to extend the capability of systems by providing intelligent feedback [383].

## 3.7 Intelligent Decision Support Systems

Any proposed Intelligent Decision Support System (IDSS) model must be designed using a MAS to provide simultaneous analysis and to enhance the fidelity of any feedback[4]. The architecture of the proposed system can resemble that of a simple thermostat, containing a monitor and feedback circuit. Building blocks of this type lead to expert systems and the creation of *production rules* in the form of logical expressions to embody *knowledge*. These rules are entered into the repository as a set of *inferences*[5] .

Therefore any proposed solution for an IDSS using a MAS to monitor and log the environment, prior to deciding on the type and amount of feedback requires significant planning. To interact, the system needs to dynamically change its input from sensor (using an event-driven model) and produce outputs to drive actuators. These agents can be instantiated using off-the-shelf expert system shells [267]. This means

---

[4] DSSs were initially generated by computer programmers in an attempt to capture the knowledge of subject matter experts in an information management system that could ideally be used to assist management in making decisions without the need of consultation or detailed analysis.

[5] MYCIN [336] and DENDRAL [105] were early commercial versions of DSSs using an expert system as its source of knowledge/inference.

**Table 3.1** Examples of Decision Making within AI

| Field | Example |
|---|---|
| Cancer Decision Support | Case-Based reasoning as a decision support system for cancer diagnosis: A case study [80]. |
| Diagnosing Breast Cancer | Using Linear Genetic Programming (LGP), Multi Expression Programming (MEP) and Gene Expression programming [170]. |
| Clinical Healthcare | Using collaborative decision making and knowledge exchange [119]. |
| Medical Decision Making | Choice of antibiotic in open heart surgery [51]. |
| Fault Diagnosis | An agent-based system for distributed fault diagnosis [318]. |
| Power Distribution | Uses Fuzzy Fuzzy Neural Networks (FNN) to predict load forecasting on power distribution networks [54]. |
| Forest Fire Prevention | Based on fuzzy modelling [167]. |
| Manufacturing | Supporting a multi-criterion decision making and multi-agent negotiation in manufacturing systems [366]. |
| Mission Planning & Security | Ubiquitous Command and Control in intelligent decision making technologies [215]. |
| Petroleum Production | Based on KBS using bioinformatics [23, 53]. |
| Production & Manufacturing | FASTCUT is a KBS to assist in optimising high speed machining to cut complex contoured surfaces so accurately that little or no finishing operation is necessary [231]. |
| PCB Inspection | Uses Evolutionary Algorithm (EA) to detect if all components have been placed correctly on the board using bioinformatics [67]. |
| Transportation | Transportation Decision Support System in agent-based environments [13]. |
| In Car Navigation | Adaptive route planning based on Genetic Algorithm (GA) and the Dijkstra search algorithm [183]. |
| Evolvable Hardware | Introduces the use of GA compile an aggregated adaptation of hardware in FPGAs [272]. |
| Detecting Spam in Email | Created an anti-spam product using a Radial Basis Function (RBF) network [177]. |
| Bankruptcy Detection | Assess an firms imbalanced dataset through the use of a classifier network [207]. |
| Robot Soccer | Using Fuzzy logic vision based system to navigate agents toward a target in real-time system [208]. |
| MAS Research Framework | Web-Based (distributed) MAS architecture to support research with reusable autonomous capabilities in a complex simulated environment [193]. |

that the knowledge needs to be represented in terms of rules generated by a SME prior to use. Such rules should be expressed in terms of *Relationships*, *Recommendations*, *Directives* and *Strategies*. Separate agents generally used to collect and refine the test data required to build and test the system. An additional interface agent (or team of agents) is used to interface the inference engine and another agent (or team of agents) to generate feedback and reports.

## 3.8  Observe-Orient-Decide-Act Loop

The OODA loop, also known as the four box method, is one approach used to aid humans when making decisions while overloaded with information. The cycle was originally labeled by Boyd as the OODA loop to assist pilots (the military decision-makers), to achieve knowledge superiority and avoids information overload, in order to win the battle[6].

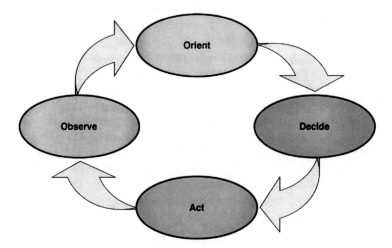

**Fig. 3.2**  Boyd's Observe-Orient-Decide-Act loop

While contemporary wisdom suggested that the win ratio of American pilots should have been better due to their training, Boyd suspected that increased losses involved additional factors. His hypothesis was that an American pilot would win almost every dogfight because he could complete loops of decision-making much faster than his adversary. Boyd constructed such a loop with the four distinct steps are displayed in Figure 3.2 [70] which could be used to assist agent learning, cooperation and collaboration. These are:

Observe:            US pilots could see their adversaries earlier and better because
                    the cockpit design of their aircraft ensured better visibility.

---

[6] Boyd studied air-to-air engagements of the Korean War (1950-1953) in which US fighter pilots, despite flying F-86 Sabre aircraft with wider turn radii had a consistent 10:1 victory ratio over MiG-15 aircraft which had much better manoeuvrability.

| Orient: | Since the adversary was acquired first, US pilots could then react by orienting themselves towards the adversary much faster. |
| Decide: | After reacting with their initial orientation, the better level of training then allowed them, as decision makers, to proceed faster to the next combat manoeuvre. |
| Act: | With the next combat manoeuvre decided upon, US pilots could then rapidly input aircraft controls, with the resultant faster initiation of a desired manoeuvre (the F-86 Sabre was more nimble than the MiG-15 because of its full hydraulic controls). |

Boyd conceptualised the principles of the OODA loop in his two famous briefings *patterns of conflict* and *a discourse on winning and losing*, which are considered the most dazzling briefings ever to come from a military mind. These began as one-hour and grew to fifteen-hour briefings over two days and were given over 1500 times. There are literally thousands of copies that have penetrated the US military and defence circles, in particular at higher levels. Boyd never formally published them but they have made him the greatest military theoretician since Sun Tzu and he has been recognised as the architect of America's strategy in the Gulf War (1990-1991) [64, 144].

The OODA loop has become a standard model of the decision-making cycle and has not only been accepted by the military, but also by many in the business community and the research community around the world [144]. It found its mark in simulation when Bratman BDI demonstrated this reasoning model [34] and the potential of becoming the method of choice for realizing truly autonomous agents. *Beliefs* represent the agent's understanding of the external world, *desires* represent the goals that it needs to achieve, and *intentions* are the courses of action that the agent has committed to follow in order to satisfy its desires [303].

## 3.9  Human-Agent Teaming

One of the major issues in early human-machine automation was a lack of focus on the human and their cognitive process. This was due to the aggressive introduction of automation as a result of urgent need. Recently, the major development in intelligent agents has become a popular choice to answer these pitfalls. Early agent models or theories are attractive solutions due to their human-like intelligence and decision-making behaviour. Existing agent models can act as stand-alone substitutes for humans and their human decision-making behaviour.

At this point we come back to one of pitfalls of early human-machine automation; the human-like substitute could fail at a critical point due to cultural diversity or lack of co-ordination leaving the human no chance of regaining control of the situation (usually as a result of impaired situation awareness). The answer to this pitfall was derived by modern AI developers who created a *machine-assistant* operating in an advisory or decision support role, that assisted human operators during critical or high workload situations. This software led to the development of intelligent agent technology. This technology has matured and is now robust enough to implement machine-assistant behaviour.

Urlings [380] claims that in order to compose effective human-agent teams with IA that include effective teams, a paradigm shift in intelligent agent development tools is required. Similarly, researchers need to transition from a technology-driven approach to automation to one that is human-centered. In his dissertation, he shows how the traditional development of agents fail to distinguish between an agent or human, preventing them from being interchangeable, even though they are *inherently different*. By establishing this difference between agents and humans, Urlings states that in a typical human-agent team, both entities are not comparable but are complementary to each other by means of cooperative sharing of task while working in concert.

This work on first principles of human centered automation is explained as follows: "Humans are responsible for outcomes in human-agent teams, the human must therefore be in command of the human-agent team. To be in command, the human must be actively involved in the team process. To remain involved, the human must be adequately informed, the human must be informed about (able to monitor) agent behavior. The activities of the agents must therefore be predictable. The agents must also be able to monitor performance of the human. Each team member (humans and agents) must have knowledge of the intent of the other members [380]."

We believe that human-centric agents could benefit from these human cognition theories as an extension of their inherent reasoning. Researchers have proved that teams can work effectively using shared mental model and R-CAST offers a promising technique for human-agent collaboration. A number of researchers in the multi-agent community are developing human-machine teaming systems for use in difficult and critical decision making under high workload situations. Human-machines teams are still led by humans but the human control will be slowly transferred to machines as the machines become autonomous and intelligent.

## 3.10 The Human Centric Approach

In order to understand where this human-agent collaboration fits into current agent trends we have to have a close look at the classification of agents. We believe that one of these classifications portrays an accurate pictures of current agent trends as reported in [280]. We propose to classify agent topology using five categorises as follows:

- Mobility (static or mobile);
- Reasoning model (deliberative or reactive);
- Ideal and primary attributes (autonomy, learning and cooperation);
- Role (information, management); and
- Hybrid (combination of the above).

The above classification philosophy forms the basis of current agent development research. Although this philosophy attempts to classify agents but it does not limit categories mixing with one another. For example the mobile agents can posses learning attribute. Here we will focus mainly on third classification based on ideal attributes as that's the leading area of current development such as teaming and learning. The third classification can assist in the development of an emerging focus area such as *collaboration* (coordination and cooperation). Nwana's system states that confusion is encountered when classifications overlap and can be labelled with greater diversity due to its combined functionality.

Purely collaborative agents are autonomous entities which coordinate their activities not necessarily collaborating with other agents (proactively collaborating activities). Collaborative learning agents are self performance improving (Learning by observation) agents by observing others (Agents or humans). The topology of an Interface agent focuses on autonomy and learning. Application areas include support or assistance to user, by adapting the agent to behave with a specific skill set, where the user 'feels' comfortable. Finally a 'Smart agent' is:

> "An agent systems that is truly smart, which would learn and react and/or interact with its external environment. In our view, agents are (or should be) disembodied segments of knowledge, embedded to represent intelligence. Though, we will not attempt to define what intelligence is, we maintain that a key attribute of any intelligent being is its ability to learn and that learning may also take the form of increased performance over time [280]."

Reasoning models of agents play an important part in their existence. They have been categorized as deliberative and reactive. Purely reactive reasoning is very much like stimulus-response type, where the action is chosen based on previously defined action-response pairs. Reactive agents are most suitable in less dynamic environments due to their responsiveness in real-time. On the other hand deliberative reasoning has inspired from cognition theories and imitates human like reasoning in agents. Deliberative reasoning is said to be slower than that of reactive but has its advantages such as more human like intelligence. This was one of the reasons why the early deliberative agent paradigms such as BDI became much popular and widely accepted in agent community.

Another major step-up in agent teaming areas is to introduce 'human-centric' nature within an agent architecture. The current trend of agent developments is concentrated on its agent only interaction, meaning that agent teaming comprised of joint goal operations that consist of agents as sole sub-ordinates of the team without any human intervention. Here we distinguish the need of human in the loop as a colleague or some times in a supervisory role. This demands agent architectures to embody social ability in order to interact with the human part of the team. Wooldridge describes the social ability as: "the ability to interact with other agents and possibly humans via some communication language [160]."

In this statement we would like to suggest that 'interaction' with humans can not only be via some communication language but also can be by other means such as observation and adaptation. We would also like to suggest that truly smart agents can be complimentary to a human by adapting similar skills as human (and that may include communication, learning and coordination) rather than being a simple replacement to the human. This leads focus to develop the agent's human centric nature by combining one or more ideal attributes such as coordination, learning and autonomy.

## 3.11 Agent-Human Coordination

The arguement of humans retaining control of any system has endured a number of centuries. While the human is expected to be in command of the human-agent team, it is also expected that an active stream of coordination flows between the human and all team members as required. Aspects of effective coordination will include control and management, communication, self-learning, performance monitoring, warning(s), and assertive behaviour.

These aspects, in descending authority relate to, task delegation and team member assertive behaviour (which is also a critical element in the human/team training concept in many training systems[7]. Complacency and changes in communication and situational awareness were factors in crew performance and human errors resulting in many incidents and accidents since the introduction of automation in the cockpit. In an analysis of the causes of accidents carried out by Boeing almost 30 years ago, it was found that human errors on the part of the cockpit crew were the primary cause in over 73 percent of the cases investigated. As a result of this initial investigation, the aviation world recognised that there was a need for an educational program to reduce errors and to increase the effectiveness of teams on the flight deck. CRM has been developed and successfully introduced in civil aviation for this purpose [397].

Urlings stipulated that since CRM has the same origin and is well-aligned with the Human Centred Automation approach, it seems logical to investigate how to translate the principles of CRM for human-human teams into requirements for constructing effective human-agent teams. It is expected that the results will not be limited to requirements and recommendations for the agent members of the team only. The principles of CRM and important skill behaviours for effective teams are well-summarised by [299] and include: "mission analysis, assertiveness, leadership, communications, situation awareness, decision making, adaptability and flexibility [380]." Where:

Mission analysis: The organisation and the development of a common plan, shared by all team-members. This includes planning, strategies or contingencies for unplanned events, task assignment and continuous task prioritising.

Assertiveness: This is a topic specifically covered by CRM programs and addresses the acceptance and expectation that all team-members raise questions when they are uncertain; or state opinions and make suggestions on decisions and procedures, even to higher ranking team-members.

---

[7] In civil aviation, originally called Cockpit Resource Management (CRM). Ten accidents, occurring between 1972 and 1982, have been cited as the motivation for creating CRM [325], while a following review of more than 35 accidents [186] provides the foundation leading to the official acceptance of CRM. Most of these accidents are also included in the research that allowed Billings [26, 27] to formulate his principles for a Human Centered Automation (HCA) approach. CRM has been a response to counter the effects of what Wiener [397] called "clumsy automation".

Leadership:            Active leadership ensures the involvement of all members in
                       the work of the team; prioritises and assigns tasks; reallo-
                       cates tasks in dynamic situations; provides clear and organ-
                       ised instructions to team-members.
Communications:        The research finding is that high performing teams apply a
                       high level of communications in abnormal situations. Infor-
                       mation in normal situations is given when required or when
                       asked, and communication is always acknowledged by a re-
                       sponse.
Situation awareness:   In effective situation awareness, potential problems are
                       identified early and other team-members are alerted. A need
                       for action will be recognised early and an attempt will be
                       made to understand the cause of the discrepancy in informa-
                       tion before proceeding. Common team awareness, instead of
                       individual awareness, is promoted.
Decision-making:       Rationales for decision-making are provided. Options are
                       considered and alternatives generated, especially in decision-
                       making for situations under stress.
Adaptability/flexibility: Plans are altered when the situation demands. Other
                       team-members are assisted when needed, and suggestions
                       and constructive criticism are accepted.

Cross-cultural differences are often discussed in relation to CRM and may have
a great impact on teaming performance and effectiveness [181]. Cultural issues are
well-defined by Hofstede [157, 158] who proposed four cultural dimensions to include the power of distance, uncertainty (avoidance), individualism and masculinity.

The discussion of cross-cultural differences are closely related to CRM. It illustrates the multi-disciplinary character and importance of social issues in effective
team building. Future powerful agents will increasingly differ in important ways from
the conventional software of the past. In order to design and build agents that are acceptable to their human counterparts, there will be a need to take into account the
social issues, no less than the technical ones [32]. The issue here concerns the Hu-
man/Agent relationship and the effect of cultural diversity may have a direct bearing
on classification and architecture. This means humans need to be considered as part
of the whole system and included in the solution. Being temporal creature, humans
can bottlenecks during Input/Output (I/O) requirements, therefore a flexible solution
is required that can maintain trust and execution without significant disruption.

## 3.12  The Next Generation

The BDI agent model has potential as a method of choice for complex reactive systems. Future trends in agent technology can be categorized on the basis of *Teaming*
which can be divided into agents, multi-agents (Teaming) and Human-Centric agent
(Human-Teaming). These two streams have two commonalities, namely, collaboration and cooperation. Along with these two commonalities the human centric agent

possesses ideal attributes such as learning. Recent work on the BDI agent such as Shared plans/Joint Intentions and JACK teams [174] facilitate agent only teaming. Furthermore the addition of the ideal attribute such as learning and collaboration enable agents to be closer to the goal of a human centric smart agent.

Agent collaboration provides the opportunity for agents to share resources during their execution. Such resources are not normally available within current MAS designs because resources are allocated for the use of specific teams. Team structures and team members are defined explicitly when the system is being designed. Using collaboration, agents are able to recognize when additional resources are needed and negotiate with other teams to obtain them. Collaboration is also a natural way to implement human-agent teaming due to the temporary and unpredictable nature of human team members.

*"Any sufficiently advanced technology is indistinguishable from magic. [248]."*

Arthur C. Clarke

# 4

# Agent System Frameworks

This chapter briefly discusses how technology evolved during the computer evolution, followed by types of agent architectures that have surfaced and how they are being used in modern systems. A brief description concludes with a discussion of future trends.

## 4.1 The Computing Evolution

The field of AI would not exist without the computer evolution. This chapter introduces the use of boolean constructs and representation, followed by a brief descriptions of the key milestones that culminated in todays computers. As researchers we need to acknowledge the science associated with manufacture of the microprocessor, magnetic and optical storage, as well as communication and transmission technologies. Other developments relating to AI are also discussed. This includes: FPGA, fuzzy logic and Reasoning, followed by a discussion of possible future direction.

## 4.2 Milestones within Computational Intelligence

There are too many events contributing to how and why computer technology evolved as it did. Although this topic is not a focus of this text, many of the principles should be revised to examine the terminology and fit, forced on the CI community. Table 4.1 lists a number of the milestones that influenced the advancement of computers since their inception. Of note is the post war contribution by Australian researchers in parallel with many of the SME pioneer nations. Australia continued to invent computers, especially in the post VLSI era and continues to collaborate on significant projects, both at home and overseas.

J.W. Tweedale & L.C. Jain: Embedded Automation in Human-Agent Environment, ALO 10, pp. 49–62.
springerlink.com                                                   © Springer-Verlag Berlin Heidelberg 2011

## 4.3   History of Logic

Using a light, battery and a specified logic gate, a specified lamp can be illuminated based on the input configuration defined by the logic table of a given circuit. The more complex the design, the more flexible the circuit functionality. An example may be to have input logic that can alter the circuit functionality based upon its state. In one configuration it may be configured as a shift register, counters, buffer, memory and I/O. The variety and range of configurations are only limited by the imagination and technical constraints.

Like all technologies, advances resulting from key events, stimulated follow-on research and subsequent enhancements or inventions. Karnaugh maps [184] (Vietch diagrams or Venn diagrams [396][4]) were all used to manipulate logic which was originally expressed in tables and Boolean expressions. DeMorgan's Law uses Venn Diagrams to assist in this transformation. Other theories available include the Pumping Lemma theory[5], MyHill-Nerode theorem[6], Quotient languages and Pseudo Theorems. Logic circuits became an instant success and formed the catalyst in the invention of the modern computer. These circuits include: *Wired-Logic, Relay Logic, Valves, Resistor-Logic, Diode-Logic, Resistor Transistor-Logic, Diode Transistor-Logic, Transistor Transistor-Logic, Emitter Coupled-Logic, and CMOS* (SSI, LSI, VLSI).

Humans don't think in binary, therefore Zadeh extended the current mathematical theories to enable programmers to use *fuzzy logic* [419]. At various stages, a number of these technologies were used to construct microprocessors and how they evolved. Number systems took a long time to evolve and its history is dependent on the country the reference author originated. The decimal system for instance, $(0-9)$[7], was also called Hindu-Arabic and is commonly called Arabic or base 10. Prior to this, the Babylonians used *sexagismal* (base 60) from 3100-2800 BC, while Egyptians and Romans[8] used letters from as early as 1000 BC[9]. Other numeral system that could be investigated include: Chinese, Hebrew and Greek. The origin of the Arabic number system is not important, the development of the *binary system* (documented by Gottfried Leibniz in the 17th century) is the major invention that significantly influenced mathematics and the computer evolution [220].

---

[1] Work started using Government Grant but abandoned in 1842 in favour of the Analytical Engine.

[2] Began work, seeded foundation work for design of the modern CPU.

[3] Based on original design by Dorr Felt invented in 1889.

[4] An example of the emergence of the Turing Machine (TM) family of microprocessors is provided in figure 4.1. The inner ring being the most dominant.

[5] Negating the application.

[6] This is the constructive aspect in its implementation.

[7] This system was heavily influenced by the Mesopotamian Indians who invented '0' as the placeholder for zero.

[8] Based on *Eruscan numerals* from ancient Rome.

[9] Labelled Roman numbers when adopted by the English in around 500 BC.

**Table 4.1** Brief History of Computer Evolution

| Year | Advancement | Pioneer | Comment |
|---|---|---|---|
| 1623 | Adding Machine | Wilhelm Schickard | |
| 1642 | Pascaline Machine | Blaise Pascal | Mechanical Calculating Clock |
| 1746 | Jacquard Loam | Joseph Jacquard | Card read weaving loom |
| 1671 | Leibniz Wheel | Gottfried Leibniz | Pascaline Machine + Multiply |
| 1822 | Difference Engine | Charles Babbage | Base$_{10}$ Design |
| 1833 | Analytical Engine | Charles Babbage | Base$_2$ Design |
| 1854 | Tabulating Machine | George Scheutz | Working Differential Engine |
| 1890 | Hollerith Tabulator | Herman Hollerith | |
| 1892 | Mechanical Calculator | William Burroughs | |
| 1906 | Themonic Valve | Le De Forest | First Vacuum Tube |
| 1919 | Flip Flop | William Eccles | Labelled coincidence circuit |
| 1924 | AND Gate | Walther Bothe | Applied to relay circuits |
| 1930 | Electronic Diff Eng | Vannevar Bush | Partial Electronic circuits |
| 1936 | Computable numbers | Alan Turing | 'Entscheidungsproblem' |
| 1938 | Z1 design $\rightarrow$ Z3 | Conrad Zuse | Used electromagnetic relays |
| 1939 | Colossus (ULTRA) | Richard Lewinski | Enigma Machine |
| 1944 | Analytical Engine | Charles Babbage | Pre-punched card design |
| 1937 | Pre-ENIAC | Heath Robinson | First used valves |
| 1939 | Harvard 'Mark 1' | Howard Aiken | Auto Sequence Cont'd Calc |
| 1945 | EDVAC | von Neumann | First transistorized CPU |
| 1946 | ENIAC | Eckert & Mauchly | First memory stored program |
| 1951 | UNIVAC | Eckert & Mauchly | Real-time with parallel logic |
| 1955 | CSIR | Pearcey & Beard | Australian/US CSIRAC |
| 1969 | CPU on a Chip | M. E. Hoff | Intel 4004 (Three chips set) |
| 1973 | First Microprocessor | Intel 8080 | Single chip with 4004 & I/O |
| 1981 | MS-DOS | Allen and Gates | IBM license the OS for the PC |
| 1982 | 16-Bit Microprocessor | Intel | 80286 |
| 1983 | Apple Lisa | Motorola 68000 | First Computer using a GUI |
| 1985 | 32-Bit Microprocessor | Intel | 80386 |
| 1985 | MS-Windows | Gates | Microsoft GUI Environment |
| 1987 | Scalar/RISC Microprocessor | Sun | SPARC |
| 1988 | 32-Bit RISC | Motorola 88000 | Use in the Apple Power PC |
| 1989 | 1 Million Transistors | Intel VLSI | 80486DX |
| 1990 | Hypertext | Berners-Lee | Stimulate the Modern Internet |
| 1993 | Pentium Microprocessor | Intel | 80586 with MMX |
| 1995 | Java introduced | Sun | Platform independent programming |
| 1997 | 802.11 (WiFi) | IEEE | Wireless standard released |
| 1997 | Cyrix | National Semiconductors | Eventually sold via TI to VIA |
| 1899 | First GPU | Nvidia | 2D or 3D graphics on Video Card |
| 2001 | SATA | International Consortium | Serial HDD Interface |
| 2004 | Ubuntu | Open source | Linux with Global language support |
| 2006 | Core CPU | Intel | Dual Core CPU (Yohan) |
| 2008 | Core 2 Duo | Intel | Quad Core CPU (Penryn) |
| 2009 | Core i7 | Intel | Multi-core series (Arrandale) |
| 2010 | Core i9 | Intel | Hex Core CPU (Clarksfield) |

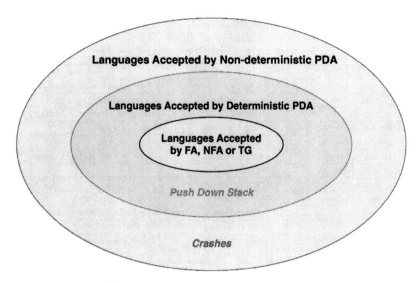

**Fig. 4.1** Venn Diagram Displaying the Evolution of the TM

## 4.4 How Computers Evolved

Computers didn't appear overnight. Like most inventions, they are the result of combining existing technologies into a single system. It is generally accepted that the efforts of many to break codes for the military during World War II fast tracked the research of this accomplishment. As many books have covered this topic a simple review of the major milestones follows:

|  |  |
|---|---|
| Abacus: | The abacus was noted in Egypt, Rome, Greece, India, and other ancient civilizations, like Mesopotamia as a rudimentary form of calculator. It evolved from a dust or sand boards (Mesopotamia), pebble line counters (Herodotus from 484-425 BC), grooved counter (Rome), the Ancient Chinese Abacus (Han Dynasty), Chinese Abacus (Ming Dynasty) and the Soroban (Japan - 1928 AD), still in use today [345]. |
| SlideRule: | Slide rules were an important phase of the computers evolution. John Napier invented the Natural logarithmic table[10], where Edmund Gunter invented the Logarithmic scale in 1620. William Oughtred quickly followed this invention with his own (side by side) two scale system in 1630 which enabled to multiplication and division [48]. |
| Babbage Engine: | The first attempt to develop a computer resulted from a government grant to Charles Babbage in 1822. He was contracted to produce a mechanical device that was designed to calculate $Base_{10}$ mathematics [12]. The *Differential Engine* |

---

[10] Henry Briggs invented Logarithms (Base 10) in 1617.

project was officially cancelled in 1833 due to the manufacturing tolerances required to achieve reliable, robust answers that could be repeated. Later that year he engaged in a new project to engineer another mechanical device he called the *Analytical Engine* [366]. George Scheutz managed to manufacture his *Tabulating Machine* in 1854 which was essentially a working version of Babbages *Differential Engine*. In 1892, William Burroughs produced the first successful Mechanical calculator which was very popular during the 19th century. It wasn't until Alan Turing published his *Entscheidungsproblem* paper that electronic computer designs started to infiltrate this market [370]. Von Neuman was instrumental in this development [389], with many other designs that emerged based on his invention. These included: Harvard, Vector, CISC and RISC designs, however these don't specifically address the architectures required to process IA or CI techniques efficiently [151].

Calculators: The calculator was invented by Kilby, Merryman and Tassel in 1967 using the Integrated Circuit (IC) Kilby invented in 1958[11] while working at Texas Instrument [38]. It contained a number of ICs, and keyboard and mechanical printer. After his move to Intel, he and Noyce[12] contributed to the first 1 Kbit Random Access Memory (RAM) chip, as series of calculator chips and the first microprocessor.

Electronic Watches: Electronic wristwatches were originally built using bipolar technology, draw substantial amounts of power and exhibit poor stability. Using Large Scale Integration (LSI) technology the tuning-fork, crystal controller and stepping motor functions can all be integrated onto a single chip. Where a Complimentary Metal Oxide Semiconductor (CMOS) substrate is used, many of the disadvantages are solved, producing viable circuits that can be mass produced. Intel developed the 5810 CMOS chip which was used in one of the first commercially available watches by Sieko .

Digital Watches: Digital watches originally used LED technology with a backlight for night use which drained the power. Over time this issue was resolved by the introduction of Liquid Crystal Display (LCD). By October 1971, 32 kHz quartz crystal controlled movements were economically viable through mass production. Following this success, Intel introduced the 4001 chip, shipped with 34 instructions and a 4-bit architecture. Its

---

[11] Seven years after this invention he dismissed his invention as *no big deal* adding it wouldn't last over the long term.

[12] Robert Noyce founded Fairchild in 1957, who produced the first commercially available IC in 1961, he also co-founded Intel with Moore in 1968 and invented the modern CPU.

primary focus is the calculator market [67]. Intel followed with a more efficient counters, calculator chips and its landmark 4004 chip set. This formed the heart the CPU, followed in quick succession by an 8-bit version called the 8008, then the 8080 and Z80 single chip CPU.

Central Processing Units: Central Processing Units amalgamated several logic functions on a single piece of silicon. It contained microcode to configure the operation of the Arithmetic Logic Unit (ALU) based on a series of binary code words. These codes have evolved to a point where they could be compiled from a series of instructions. Modern CPUs have a rich instruction set that enables it to complete complex programs.

Personal Computers: Personal computers may contain multiple CPUs with multiple cores. Tape machines have given way to secondary storage and screens now display text or graphics as the result of the program execution. Speed and parallel cores have dramatically evolved over the past 60 years, however the design and flexibility of CPUs require further development to assist developers in producing applications or systems using intelligent agents.

Recursion: Recursion is typically a three step process: the basic operation (algorithm), rules to create objects, and object (scope or declaration i.e. given every $X^{th}$ object). The symbols use include the basic Arithmetic Expression (Æ), the negation operator[13], typically expressed as ¬. Another expression derived during this era was the Well Formed Formula (WFF). It is based on the Æ that includes a series of legal syntax expressions, such as; $\Sigma = \{ \neg \rightarrow ( ) a\, b\, c\, d \dots \}$ which can be formed using the following rules:

- The expression contains any single Latin letter a b c d ... ,
- If $p$ is a WFF, then so are $(p)$ and $\neg p$, and
- If $p$ and $q$ are WFFs, then so is $p \rightarrow q$.

Storage: The Winchester Hard Disk Drive (HDD) originated from IBM with less than 1MB of storage. This capacity and speed rapidly advanced with a variety of devices now on the market. Solid State Drivess (SSDs) are manufactured using silicon memory. In the past they have used NOT AND (NAND) memory technology, however Ferromagnetic Random Access Memory (FRAM) devices are beginning to appear. The single biggest problem with SSD is speed, followed quickly

---

[13] A developer could also expression negation as an: *'invneg'* which is the reverse of the negate operator, '−' the raised bar and '∼' the tilde operators.

by capacity and price. Samsung provided a SATA II interface and improved the specifications with the introduction of
a 128 GB 1.5 or 2.5 inch Multi-Level Cell (MLC) drive with
100MB/s read and 70 MB/s write speeds [75]. The power
consumption during use has also been reduced to 0.5W. Samsung released a 256 GB MLC SSD, with read and write
speeds of 200 MB/s and 160 MB/s respectively. This device
will consume 0.9W of power when active [76]. They are 2.4
times faster than the typical HDDs, but they cost more per
GB. Seagate has since introduced a hybrid drive with hugh
SSD cache (4GB) and 500 GB of traditional storage. Since
introducing the 1 TB drive in 2007 and 2 TB in 2009, the 3
TB HDD is now overcoming the FAT16 address restrictions
[77]. Hitachi forecasts that it will be capable of manufacturing a 25 TB HDD by 2018 and plans to introduce a 1TB SSD
shortly [332].

Compilers:                Compilers have evolved significantly to a point where real-
time compilation is being the norm (Java compilers pro-
duce byte code that is interpreted in real-time). These techniques were borrowed from advances in horizontal architectures like Advanced Flexible Processor (AFP) designed by
Control Data Corporation (CDC) and other multi-pipelined
vector processors. Code generation is one of synthesis and
scheduling [310], where task decomposition is required to
assist in resource allocation and constraint minimization[14].

Computer Designs:   The Turing Machine (TM) was created using the theory of
a deterministic machine with an output model [369]. His design was enhanced after a number of modifications by Post
and Minsky. Post included the concept of an embedded execute cycle that contained *Halt, Accept and Reject* commands.
The *Minsky Machine* consists of a two-stack PDA. There is
a natural transition from Finite Automation (FA) to PDA to
TM. Other variations of the TM display the separation of
the program(s) and the application. Part of the TM can be
re-written based on a cross between the Meeley and Moore
machine in 1955[15].

It is the abilities of many pioneers that led to the von Neumann CPU design that was able to *Add, Subtract, Multiply
and Divide*. Subsequent researchers enhanced these concepts
to contribute to the modern era of computing. Some of those
milestones are introduced in the following text.

---

[14] This is required in order to create code that programmers can comprehend and maintain.
[15] Alonzo Church's Theorized in 1936 that a function can be computed by a well written
algorithm in a TM.

## 4.5    Types of Architecture

An Agent Architecture is considered to include at least one agent that is independent or a reactive/proactive entity and conceptually contains functions required for perception, reasoning and decision. It may also be viewed as a particular methodology for collectively connecting agents or agent sub-systems. The architecture specifies how the various parts of an agent can be assembled to accomplish a specific set of actions to achieve the systems goals.

The next major enabler required to create a new generation of agents is to incorporate *human-centric* functionality. The current evolution of agent systems has concentrated solely on agent-to-agent interaction and fail the "social ability test primarily because they are unable to *interact* with other agents via a common language [150, 410]". A multi-purpose language will benefit AI research, when it conforms to unified standards and frameworks, using the underlying protocols. Interaction with humans is not limited to a communication language but should be extended to include observation and adaptation. The suggestion is that smart agents work in conjunction with humans by adapting or translating their skills (communication, learning and coordination) rather than the functions they conduct. It should be noted that components have successfully been enhancing the dynamic nature of agents. These concepts help focus our development of agents adopting a human centric nature by merging one or more of the ideal attributes used in coordination, learning and autonomy.

Forming agent teams generally requires prior knowledge of the resources and skills required by teams to complete a goal or set of decomposed tasks. This will be determined by the maturity and composition of the team, its members and any associated team[16]. The maturity could be measured across each phase of *Team Development'* as described by Tuckman [238].

Personified characteristics and functions are possible in BDI agent architectures, although each function consumes, duplicates or retains resources[17]. To reduce overheads, only the resources or functionality required to process the task need be instantiated and released after the function is no longer required. Based on this premise, interactions between agents within teams and between teams can generally be catalogued.

## 4.6    Existing Architectures

An exhaustive search on how the definition for an agent was derived [116], where Bratman [34], Russel and Norvig [320], Jennings and Wooldridge [176] all contributed to enhancing this definition, although many of the links need to be enunciated. The features discussed include: autonomy (decisions/actions), reactivity (perception, behaviour or beliefs), pro-activeness (purposefulness/goals/ intentions) and

---

[16] Each Agent may interact with other agents or group of agents (a Team). That agent may also interact or collaborate with other systems, within or across multiple environments. Agents can collaborate freely with other agents, although *Teams* are forced to communicate using traditional hierarchical methodologies. This structure needs to become more flexible in order to deliver efficiencies.

[17] Many once or sporadically during the life of the application.

social ability (ability to personify, communicate/coordinate/ collaborate, or inter-act). Each has been debated, for instance Frankcik and Fabian [113] classify gestures (written or spoken) as the ability to stimulate or *act*, Reynolds [312] and Tu [368] compare perception with the paradigm of *behaviour* and Bratman [34] attempts to tie this together into an architecture called BDI. Evans [98] labels this approach as agency *orthodoxy of agent technology*, where intentions can be mapped to *opinions* and actions (re-actions) a consequence of *intentions* as implemented by Rao and Georgeff [303] in their BDI architecture.

A real-world system takes inputs as sensors and react appropriately by modifying the outputs as necessary. Simulation models rely on the same approach. They monitor sensors that stimulate decision making that may cause changes to outputs. Three architectures have been dominant in AI research: blackboard systems, contract nets and frameworks. A blackboard may be thought of as a componentised system, where each box could function as separate concepts that represent the specified aspects of a system or sub-systems engaging the problem. This happens in an environment where experts are modular software subsystems, called knowledge sources, that are capable of representing different points of view, strategies, and knowledge formats, required to solve a problem. These problem-solving paradigms include: Bayesian networks, genetic algorithms, rule-based systems, case-based reasoning, neural networks, fuzzy logic, knowledge-based systems, legacy (traditional procedural) software and hybrid systems.

## 4.7 Architecture Limitations

The largest impediment to any application is security. The Java security manager in the JVM has been specifically designed to protect against malicious or errant applets. By limiting communications to a specific range of sockets, security can be further enhanced. The use of dedicated protocols and/or event isolates components semantically to an even higher-level. Finally using packaging in the form of JAR files, *mobility* is enabled and using sophisticated reasoning and learning, while architectures that support *autonomy* are becoming mainstream and easy to implement [25].

Two types of search have developed; brute-force and heuristic. Where brute-force algorithms conduct blind searches that examine every node of the state space for a solution and heuristic algorithms conduct informed or directed searches based on variables bound to the game state to improve the efficiency of the search. The best solutions are considered complete when they can be solved within minimal time and space. An agent architecture is considered to include at least one agent that specifies how the various parts of the framework can be assembled to accomplish a specific set of actions. Some researchers classify agents by their functionality, while others used utility or topology [86]. Pressure is emanating from within the DAI community to include interaction within the BDI agent paradigm.

A multi-agent system may be regarded as a group of ideal, rational, real-time agents interacting with one another to collectively achieve their goals. Each one of these individual agents needs to be equipped to reason not only about the environment, but also about the behaviour of other agents. In this context, a strategy refers to a decision-making mechanism that provides a long term consideration for selecting actions to achieve specific goals. Each strategy differs in the way it tackles the solution space of the problem.

## 4.8    Integrated Systems

Data and information processing paradigms that exhibit the main attributes of intelligence could be referred to as members of the family of techniques that form part of the knowledge-based engineering domain. Researcher have enabled AI developers to create 'Human-like' applications [171, 379] using agent-oriented paradigms. Relevant examples of existing research for: Knowledge-based systems [41, 321], Blackboard systems [278], Model-based reasoning systems [337], Artificial neural networks [172], Evolutionary computing [317, 390], Fuzzy logic systems [173, 326], Hybrid systems [171, 380] and now BDI based agent-oriented systems [87].

## 4.9    Steps towards Next Generation

The BDI agent model has potential as a method of choice for complex reactive systems. Future trends in agent technology can be categorized on the basis of *Teaming*. Agents, multi-agent (Teams) and Human-Centric Agent (Human-Teams) and hybrid systems should all be considered. These streams have two common traits (collaboration and cooperation). Along with these two commonalities the human centric agent possesses ideal attributes such as learning. Recent work on the BDI agent such as Shared plans/Joint Intentions and JACK teams [6] facilitates agent only teaming. Furthermore the addition of the ideal attribute such as learning and collaboration enable agents to be closer to the goal of a human centric smart agent.

Agent collaboration provides the opportunity for agents to share resources during their execution. Such resources are not normally available within current MAS designs because resources are allocated for the use of specific teams. Team structures and team members are defined explicitly when the system is being designed. Using collaboration, agents are able to recognize when additional resources are needed and negotiate with other teams to obtain them. Collaboration is also a natural way to implement human-agent teaming due to the temporary and unpredictable nature of human team members.

## 4.10    Dynamic Agent Capability

Currently agents represent intelligent spiders, bots and aggregators [149] that use HyperText Transfer Protocol (HTTP) to pull data from the web before filtering the content[18]. Some sites, such as Amazon and E-bay, collect persistent data about its users to enable the pages presented to become personalized. Many overlook the push technology [358] provided by Corba, Java Beans and .NET[19]. These are event driven

---

[18] Search agents and filter agents have since been incorporated into domain specific assistants, usually in the form of an information or interface agent. These are predominantly based on learning, either by observation, advice, reinforcement or collaboration [234].

[19] Including OLE, ActiveX, COM and DCOM from Microsoft.

and are able to plan or initiate an action autonomously based on sensors to detect any changes in the real-world situation. Running agents unsupervised takes a leap of faith, hence many agents are still passive in nature [374].

MAS development consists of few expert agents and multiple client agents. These agents have common functionality that is supported with some common capabilities that are predefined and one or more specific capabilities that are inherited (possibly dynamically). The expert agents are equipped with specific capabilities to assist other agents in resolving their problems. Once the expert agents are created and initiated they would stay active for the lifetime of the program waiting for client agents interaction. In an experiment, client agents were initiated from a single java class and set to roam the problem space. The technology is available to assign different actions and tasks. These actions and tasks are defined by the operator when creating the agent. For example, action of client agent could be set to translate and the task could be any text that user wishes to convert. The other available actions attempted in this program are to calculate and parse into Extensible Markup Language (XML) that can be selected as the goal of the client agent. The mathematic expression to be calculated or the code that is to be parsed into XML would be entered in the task field. According to the required action of client agents, the expert agents are introduced to client agents by the facilitator agent. Three different expert agents are developed in this system. The first expert agent is called translator because its expertise is translating an English word or phrase to its other English synonyms. The translator agent has access to a dictionary database that has the meaning of English words. The second expert agent is called calculator because it calculates the result of short expressions that client agents would ask them. It calculates the four common operations of summation, deduction, division, and multiplication. The mathematical expression is entered as a string by the user when creating the client agent. The calculator agent then splits the numbers and calculates them based on the mathematic operator in between them. Therefore, the format of the expression has to be specifically entered as 'A * B' in order for expert agent to understand and perform the calculation. The concept of the relationships for an agent factory is shown in Figure 4.2.

## 4.11 The Dynamic Architecture

The maturity could be measured across each phase of '*Team Development*' as described by Tuckman [238]. Personified characteristics/functions are possible in BDI agent architectures, although each consumes resources that may only be used by relatively few agents, many sporadically during its life. To reduce these overheads, only the resources/functionality required for a specified period would be instantiated and then released some time after that function is no longer required (Similar to the concept depicted in Figure 7.3). Based on this premise, interactions between agents within teams and between teams can generally be catalogued. We expect that IA will retain their architectural foundations, but that the availability of more appropriate reasoning models and better design methodologies will assist them in being used increasingly in mainstream software development. Furthermore, better support for human-agent teams will see the development of a new class of intelligent decision

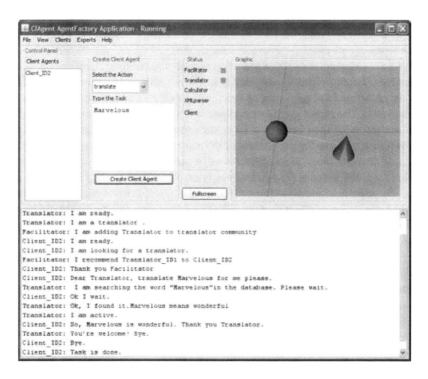

**Fig. 4.2** Snap Shot of the Agent Factory Demonstrator

support applications. Autonomous connectivity using one of the existing communication modes will be required to assist with the higher level interoperability.

Establishing communication between heterogeneous agent systems is crucial for building trust relationships between agent instances in these systems. Interchanging messages between different agent systems situated in different environments possibly affects all layers of the communication model, including transport protocol/system ports, ACL, semantic layers and ontologies. The most common languages currently include: KIF, ACL, KQML and FIPA/ACL.

Incoming communication requests can be forked to autonomous threads capable of handling specific protocols and further processing and translation. To build a *proof of concept* for the communication layer of the TNC model we implemented a multi threaded client/server system to simulate the processing of multiple signal sources. These input stimuli are described by clients connecting to a server using a random protocol in our function model. The server thread models the communication interface of the receiving agent. The servers master thread accepts connection and forks it into a new thread specific for the determined protocol type, the received information is processed and committed to the knowledge base in the joint step. The communication prototype is written in Java using Sun's$^{TM}$ system library for multi-threading and networking. The function model is a simple Java application implementing the above architecture and visualizing the on-going communication.

## 4.12 Open Architecture

Many systems fielded in the military have designs (using closed architectures) that are over a decade old before being commissioned into service. They use proprietary software that is often tied into the existing hardware and rely heavily on technical refreshers to implement new capability (technical insertions). Five challenges have been identified to address the move towards open business and architectural models. The first is to clearly *delineate* the software from the hardware. Next is to force project managers to *use the Advanced Processing Build (APB)* methodology, and avoid obsolescence by planing for *technical insertions* based on Moore's law [257]. The other two challenges call on the premise *"design once, uses many times"* and linking the project to an operational or capability map [350].

## 4.13 Future Architectures

For the formation of a dynamic architecture that behaves with *Plug & Play* flexibility retains the power of embedded business logic with the flexibility of context switchable scenario data. Using a common interface with reflective components will provide minimal overheads to achieve the maximum efficiency available in a distributed environment. Establishing and implementing this process across a finite set of protocols will constitute the first phase of the implementation of the model. These agents must have the capability to autonomously context switch within a teaming environment to successfully achieve task/resource management in an MAS. Cohen [58] describes a Situation Specific Trust (SST) model comprised of a qualitative (informal core) that describes the mental attributes and includes a quantitative (prescriptive extension). Using taxonomy as described it is easy to extend the capability of any system dynamically to solve both descriptive and ad-hoc problems. A concept demonstrator is being developed to demonstrate this research and enable further study into the dynamic operation of agent architectures, capable of auto negotiating with other agents, systems or sub-systems.

*"The economy of human time is the next advantage of machinery in manufactures [12]."*

Charles Babbage

# 5

# Agent Interoperability and Adaptation

In this chapter a brief description of syntax, semantics, topology, ontology and taxonomy is discussed in relation to agency theory. The Open Systems Interconnection (OSI) model is included to help aid the ready align the technical, physical and cognitive connections. At present the communications layer is being embedded into many agent architectures, when in fact, it should remain as an underlying transport mechanism. However the interoperable functions are still required as are a standard methodology of exchanging information and intent. These and related subjects mechanisms are also included to generate further discussion into what the future standard should include.

## 5.1 Syntax and Semantics

The Syntax of a language is driven by its vocabulary, while the semantics is structured to ignore the ambiguities resulting from divergent human interpretation. The term was introduced by Bréal in 1897 [35] in conjunction with the literal interpretation of natural language. An example commonly quoted is: *time flies like an arrow*. For instance, some people interpret *time* as speed or *flies* as the ability of an arrow to fly straight or true. Programmers need to delineate between these concepts, introduced by metaphors and apply a distinct, less complex form of truth conditions (possibly via a series of lexeme) [142, 294]. Logical and procedural languages both include semantics, the first being to correctly describe the software algorithms or functions while the later is used to generate semantic-webs or maps. These topics are well covered in many texts, however as it forms an important role in communications, compliance is implicit especially with an alternate context.

J.W. Tweedale & L.C. Jain: Embedded Automation in Human-Agent Environment, ALO 10, pp. 63–72.
springerlink.com

## 5.2   System Topology

The concept of a topology relies on the mathematical principles of spacial presence or connectivity, using a compact design. The term was introduced by Listing in 1847 with respect to geometry [226]. Cantor later used this term to describe point sets within *Euclidean Space* when calculating *Fourier* series [137, 179] and Wolfram to describe deformation of objects [404]. Due to the numerous definitions of this term, our focus is on the layout or structure of systems, such as; linear, star, ring and tree network topologies.

## 5.3   Data Ontology

It is generally acknowledged that descriptive languages have been the most used tool to describe what is known about a system or domain [140]. This argument revolves around *an explicit specification* of how a system is *conceptualised*, using the declarative nature of the description to understand its complexity or the semantics of the language[1] used to model it.

The term ontology has been loosely used to describe many relationships, however during this review it is said to be "initiated in AI, use common-sense knowledge that is accessible and can be processed. This knowledge is grouped into concepts or micro theories that express the context dependency of the knowledge and who's upper levels are publicly available [314]". This is demonstrated using Knowledge Interchange Format (KIF) in Grubbers' famous elevator example, which forms an "object, concept and controlling entities that are presumed to exist within the relationships that binds them. [127]

## 5.4   Taxonomy

Agent Taxonomy is formally viewed on a number of planes. These include intelligence[2], mobility and autonomy[3]. Some of the processing strategies include: *reactive* (reflex [36]), *deliberative* (goal directed [265]) and *collaborative* (BDI [34]) frameworks. Lange suggested the following as the most desirable attributes [213]:

- dynamically adapt to load changes,
- encapsulate protocols,
- fault-tolerance,
- heterogeneous,
- operate autonomously,
- overcome network latency, and
- reduce network load.

---

[1] Business Process Model (BPM), Business Process Execution Language (BPEL), OWL-Services (OWL-S), Unified Modelling Language (UML), Petri-Net Markup Language (PNML), DARPA Agent Markup Language - Services (DAML-S), and even the Web-Services Description Language (WSDL).

[2] More formerly termed agency.

[3] This includes the level of functionality performed independent of human supervision.

## 5.5   Open Systems Interconnection Reference Model

The International Standards Organization published the seven layered Open Systems
Interconnection Reference Model (OSIRM) in 1982 to reduce the plethora of diverse
communication designs emerging in applications that were destined to become or-
phans. Several companies took advantage of the chaos attempting to create proprietary interfaces or specialize in writing sockets [168]. These layers are shown in Figure 5.1 (modified to illustrate the fit) and include:

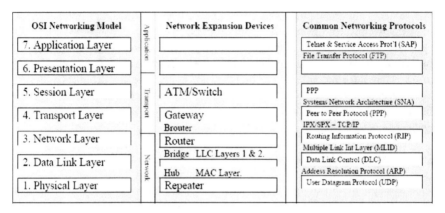

**Fig. 5.1** Rationalized OSI Reference Model [168]

Physical Layer 1:   The means of transmitting the data may include Optical, Mechanical, Electrical media; such as, Accoustic, Radio, Op-
tical and Networks derived mechanisms[4].

Data-link Layer 2:  HDLC, 802.11 (WLN), ATM, PPP.

Network/Internet Layer 3:  IPV4 (RFC 3927), IPV6, NCP, X.25.

Transport Layer 4:  TCP, UDP, WAP Datagram Protocol[5].

Session Layer 5:

---

[4] The Physical protocol is a set of rules that govern how to communicate between entities. generally restricted to wire based transmissions, such as, User Datagram Protocol
(UDP) [295], File Transfer Protocol (FTP) [297], Transmission Control Protocol / Internet
Protocol (TCP/IP) [296] and Universal Plug and Play (UPnP) [218]. In human-machine
systems a hybrid form of communications is required. This is especially important when
equipment is used in mobile or remote locations. The means of transmitting the data may
include Optical, Mechanical, Electrical media; such as, Accoustic, Radio, Optical or simply
Network derived mechanisms.

[5] Communication is achieved across a number of layers. The layers depend on the Language
and Protocol(s) used. ACL [339], Foundation of Intelligent Physical Agents (FIPA) [283],
Knowledge Query Manipulation Language (KQML) [108], SOAP [97].

Presentation Layer 6:
Application Layer 7: DHCP, FTP, HTTP, POP3, SOAP, NTP (FFC 867-8)[6&7].

Little has been done to represent a uniform model that embodies connectivity at or above the session layer. With most network based solutions, the physical layers are ignored, as the standards are already embodied in commercial hardware. As business activities become more autonomous and/or distributed, researchers are required to experiment with the layers above the predefined standards embodied within ISO reference model. Communication between pre-existing links that contain those disparate standards can only be forced to achieve interoperability and successfully exchange data between applications. The sequence begins in the OSI reference model and extends upwards towards the cognitive plane.

For the purpose of this text three categories are used to achieve many of the above attributes. These include the physical data flows that would include control (*DIS, HLA*), where information (intent), command and control (desire) is treated as the primary intermediate flow (*ACL, KQML, FIPA, SOAP*) and cognitive knowledge (beliefs) at or above the OSIRM layers in a manner that is intended to achieve self-synchronous behavior.

## 5.6 Agent Communication

Successful communication requires shared knowledge and a common communication language. It is important for agents to retain knowledge about their environment to enable them to conduct intelligent actions. The meaning of similar concepts or entities must be the same among agents even if they use different names. This criteria leads to definition of shared knowledge of concepts in each particular domain of knowledge that is known as ontology.

In 1990, Defense Advanced Research Projects Agency (DARPA) of US Defence Department initiated the Knowledge Sharing Effort (KSE) in order to develop knowledge sharing techniques and tools [209]. As part of KSE, KIF was proposed as a logic language standard to describe things within computer systems and help interchanging knowledge between agents in order to improve their interoperability.

We expect that AI will retain their architectural foundations, but that the availability of more appropriate reasoning models and better design methodologies will assist them in being used increasingly in mainstream software development. Furthermore, better support for human-agent teams will see the development of a new class of intelligent decision support applications. Autonomous connectivity using one of the existing communication modes will be required to assist with the higher level interoperability. The most common languages currently include:

---

[6] Software semantics are also enforced across a number of layers. Again these are language dependent. Standard Generalized Markup Language (SGML) [63, 395], HyperText Markup Language (HTML) [302], Dynamic HTML (DHTML) [134], Extensible HyperText Markup Language (XHTML) [291], XML [322].

[7] The means by which the signal is encoded and decodes. Optical, Mechanical and Electrical signals may all be modulated prior to transmission. The general classifications include, amplitude, frequency/phase and pulse modulation.

- Agent Communication Languages (ACL),
- FIPA Agent Communication Languages (FIPA ACL)[8],
- Knowledge Interchange Format (KIF), and
- Knowledge Query Manipulation Language (KQML).

The interaction and cooperation of agents in MAS is not possible without some kind of ACL. Austin was one of the pioneer researchers in the area of ACL since 1962 and his work inspired other researchers in this field. He stated that communication exchange contains the intention of speaker and speakers perform speech actions when stating certain class of utterances, which he referred them as performatives (also known as acts) [10]. Performatives can produce actions or affect feelings, thoughts and beliefs rather than just stating facts. Some examples of performatives include suggestion, warning, request, and so on. Searle continued Austin's work in 1969 and formalized the structure of speech acts by defining their different classes and the necessary and sufficient conditions for their successful performance [330]. Cohen and Levesque introduced speech acts in AI. They developed plan-based theory of speech acts and showed how pre/post-conditions of speech acts could be represented in a multi-modal logic [60, 62] .

Using ACL, agents can follow shared sequence of messages related to a particular topic or task and negotiate for their needs, preferences and constraints. ACL messages are usually defined in terms of the BDI model [209].

## 5.6.1 Agent Communication Language

The knowledge of agents about their environment is important for them to take intelligent actions [25]. Agents can be equipped with sensors to obtain information. Communication of agents can be directly between two agents or through a facilitator or interpreter. Communication and exchange of information in any domain requires each agent to access shared knowledge within that environment is known as ontology. Each domain of knowledge normally has its own ontology.

A successful communication requires shared knowledge and a common communication language. Agents communicate by sending messages to each other. It is required that the format and semantic of theses messages be accepted by all the agents in the MAS [385]. Therefore, standards are required to be accepted in order to establish compatible communication among the agents in open MAS environment. These standards are defined in terms of ACL.

Communication has three aspects: syntax, which is the structure of communication symbols, semantics, which defines the meaning of the symbols, and pragmatics, which provide the interpretations of the symbols. The combination of these produces meaningful phrases for agents to communicate [103]. The semantic content of messages should be consistent for all agents even if they use different names [209]. Therefore, Agent Communication Languages (ACLs) were developed to handle complex

---

[8] As this standard matures it has split into representation and implementation models, DARPA Agent Markup Language - Services (DAML-S) and Web-Services Description Language (WSDL).

semantic and transfer messages that describe a desired state[9]. They also enable the exchange of more complex objects such as shared plans, goals, experience, and so on [209].

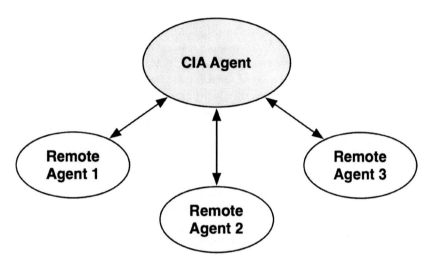

**Fig. 5.2** Agent Factory Demonstrator Framework

Rasmussen labels knowledge as the combination of facts, conditions, expertise, ability or understanding associated with the sensors (sight, taste, touch, smell and sound) relating to anything in their environment [305]. This originally constrained the analysis and processing of knowledge to symbolic representation [306, 307, 308, 309]. Early systems were forced to store symbolic tables in flat databases, however the growth in capability of hierarchical, relational databases has extended the scope of knowledge engineering, especially in expert systems [54, 382].

Agent communication languages have been used for years in proprietary MAS. Yet agents from different vendors or even different research projects cannot communicate with each other. Components will be added dynamically and will be autonomous (serve different users or providers and fulfill different goals) and heterogeneous (be built in different ways) in MAS. To fill the concept of interoperability and autonomy [348] agents must be able to talk to each other to decide what information to retrieve or what physical action to take, such as shutting down an assembly line or avoiding a collision with another robot. The mechanism for this exchange is the agent communication language. KIF can be used to extend the simple transaction to pass information into one that share knowledge using a standard interchange format.

### 5.6.2 Knowledge Interchange Format

The KIF language [320] was designed as a solution to the translational problem. It is a logic language used as a mediator to translate between two disparate languages.

---

[9] JATLite and the JAFMAS are agent communication architecture [373].

KIF is a prefix for first order predicate calculus with extensions that support non-monotonic reasoning and definitions. The language description includes both a specification for its syntax and one for its semantics.

The KIF language [320] was designed as a solution to simplify the problem of translating between disparate models. It is a logic language used to describe things within computer systems such as expert systems, intelligent agents and so on. KIF is a prefix version of first order predicate calculus with extensions to support non-monotonic reasoning and definitions. The language description includes both a specification for its syntax and one for its semantics.

### 5.6.3 Knowledge Query Manipulation Language

The first inter-project ACL was KQML[10] queries in KQML referred to as performatives, a term adapted from speech theory[11]. The syntax of KQML statements are a subset of LISP grammar, although parameters in performatives are indexed by keywords rather than position. A statement is composed of a reserved performative keyword. Parameters contain information on the communication act itself and the contents of the message (these parameters represent the language and ontology).

KQML is a high-level, message-oriented communication language. It is based on speech act theory and consists of primitives which allow agents to communicate their attitudes. KQML is independent of transport mechanism, content language, and ontology [209]. The syntax of KQML is based on the Lisp language.

KQML has three layers of content, message and communication [103]. Similar to the KQML language, FIPA is also based on the speech act theory and their syntax is similar except for some reserved names of the primitives [209]. In FIPA ACL, primitives are called *Communicative Acts* (CAs) but they refer to the same entities as primitives in KQML. Foundation for Intelligent Physical Agents, FIPA ACL has played an important role in defining agent communication standards [283]. Its aim is to promote agent-based technology and produce specifications on ACL [209].

### 5.6.4 FIPA ACL

In the early 1990s, France Telecom developed *Arcol* which includes a smaller set of primitives than KQML . The primitives are all 'assertives' or 'directive's', but unlike KQML they can be composed. *Arcol* has a formal semantics, which presumes that agents have beliefs and intentions, and can represent their uncertainty about various facts. Arcol gives performance conditions, which define when an agent may perform a specific communication.

The core CAs are inform and request and other CAs include inform-if, inform-ref, request-when, request-whenever, agree, cancel, confirm, refuse, propose, query-if [103]. The messages in FIPA ACL are considered to be CAs and are the compulsory parameter of the message [373] .

---

[10] US Defense Advanced Research Projects Agency's Knowledge-Sharing effort in the late 1980's [107].

[11] An utterance is a *performative* statement when its function is to perform the action mentioned as apposed to a propositional express about it.

The agents communication is not just about exchanging bits of data but it includes communication of complex attitudes or behaviours [209]. The agents might inform, request, or negotiate in performing some task. Communication topology defines which agents are allowed to communicate with each other [90]. For example, in some cases, agents can only broadcast messages to all other agents. However, it is not always desirable that all agents receive the same message; therefore some mechanisms are defined to structure agents communication.

### 5.6.5    FIPA versus KQML

FIPA ACL is superficially similar to KQML. Its syntax is identical to that used by KQML , except for the different names for some reserved primitives. KQML separates the outer language that defines the intended meaning of the message and the inner language, or content language that denotes the expression represented by the beliefs, desires and intentions of the interlocutors, as described by the meaning of the communication primitive. The FIPA ACL specification document claims that FIPA ACL (like KQML) does not make any commitment to a particular content language. Although such a claim holds true for most primitives, there are FIPA ACL primitives for which some understanding of the language Semantic Language (SL) is necessary for the receiving agent to understand and process the primitive [207, 208]. Agent communication architectures apart from KQML and FIPA ACL are either ad hoc JATLite, JAFMAS or not fully documented as in proprietary systems.

### 5.6.6    SOAP

Simple Object Access Protocol (SOAP) is an inter-application cross-platform communication mechanism. It defines a simple and flexible communication format that is based on XML [331]. SOAP is highly extendible across different programming languages, platforms, and operating systems as long as they can generate and process XML. It increases the interoperability of distributed systems such as the World Wide Web.

Using the SOAP mechanism, information is packaged in SOAP messages that consists of two main elements of envelope and body, and optional header elements [136]. The envelope element contains attributes such as the encoding style attributes that specifies the encoding style of the message. The body of message includes the information intended for recipient. This information should use the SOAP encoding rules. The header element can be used to add information about the message such as the return path for the response.

- Different strategies have been developed to enable the communication of software components and the exchange of information and knowledge between applications [331].
- Some examples of these techniques includes: Remote Procedure Call (RPC), Remote Method Invocation (RMI) in Java, Distributed Component Object Model (DCOM), and Common Object Request Broker Architecture (Corba).

- The CORBA enables the communication between software components and procedures that are written in different languages on different computers [209]. Its messages only include procedures and they do not contain any semantics.
- The Web is the largest RPC mechanism where objects can be requested and accessed by many computers around the world.
- Some standards for distributed computing includes: Distributed Computing Environment (DCE), DCOM, and Corba.
- The DCOM provides an object-based RPC mechanism. and
- SOAP specifies three items: envelope, encoding rules, and RPC representation.

### 5.6.7 Extensible Markup Language

The XML is used to represent data in a standardized text-based format [21] and sharing it across different platforms. XML makes data transportable to other systems. Structured data can be extended by XML tags. The data validation Document Type Definition (DTD) defines ontologies in XML. The XML Parser agent is another expert agent that is developed to parse the XML data into HTML code.

### 5.6.8 DARPA Agent Markup Language - Services

The functional description of DAML-S uses WSDL to represent the service profile using Inputs, Outputs, Preconditions and Effects (IOPE) and the service model describes the internal Process (Meta) Model. WSDL is used to document the abstract interface grouped as ports (Meta-Models) and with the concrete implementation of the protocol using SOAP to communicate via a HTTP format.

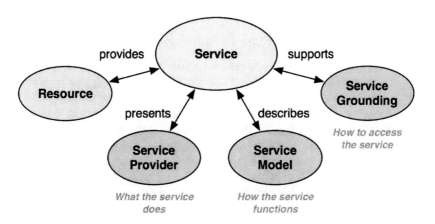

**Fig. 5.3** Top Level of the Service Ontology [239]

DAML-S[12] has a pedigree that is some what schizophrenic. Over time, opposing paradigms have progressively attempted to defend disparate styles of describing

---

[12] Web Services = DAML+OIL Ontology.

semantic representation using real world problems. WSDL is one of the languages used to describe and invoke web-services, however like BPEL, Ontology Inference Language (OIL) and KIF was quickly replaced by DAML-S/BAML-OWL (DAML-W) [140] . Using DAML-S, where DAML + OIL represents Ontology Web Language (OWL), ontology presents the what, describes the why and supports the how to, as shown in Figure 5.3. Unlike WSDL, there are very few examples on how to use DAML-S, although some have been written to demonstrate how it provides an imprecise conceptual model (as an isolated technology). As such, it is difficult to learn. WSDL does provide a limited form of expression but it can map to DAML-S and remains superior for at least the top few layers. Therefore future use of DAML-S is recommended for large scale usage and the realization of a semantic web for information integration.

### 5.6.9   Web-Services Description Language

Although this topic has a strong following in the web-based application fraternity, there are major issues with corporate level applications because of distribution and security. Like many other aspects required to conduct autonomous dynamic reconfigurable agent teaming, much needs to be improved. However many of the issues parallel to those experienced when attempting to conduct distributed communications. Web services use a number of XML technologies in the SOA environment. It uses posted and web document style messaging. Parameters used may be serialized and return values in RPC and XML protocols. The latest standard supports *metadata* to describe the interface being used to bind the communications protocol, port, or service Universal Resource Locator (URL), providing a powerful solution to interoperate with *loosely coupled, coarse-grained, distributed* SOA applications [316].

### 5.6.10   Future Communications Standards

Regardless of the protocol used, agents require different forms of information, knowledge and context which all have temporal elements. The context generally describes the state of the environment, where events can force changes to occur rapidly. An example could include navigating through an unknown environment (considered friendly) until a shot rings out. Many flow-on actions may occur, such as, a flight or fright reaction, switching to survival mode or simple health monitoring. Game manufacturers have already recognized that *static* and *dynamic* communications channels are required. The same is experienced in modern simulators and some real-world systems. Like hardware, communications mechanisms need to evolve further.

*"Despite all its success, there is still much that goes on in nature that seems more complex and sophisticated than anything technology has ever been able to produce [407]."*

Stephen Wolfram

# 6

# Enhancing Autonomy

The question of autonomy using agents is complex and requires significant skill to write applications that interoperate successfully. The complexity is derived from the need to cater for every conceivable outcome. Providing applications with the ability of being able to adapt dynamically is vital in this domain. A brief discussion of design patterns, threads, components and associated architectural research are used to assist the reader in understanding this topic.

## 6.1 Dynamic Agent Components and Communications

Dynamic agents and flexible communication models all exist, however non are efficient or flexible enough to solve large scale real-world problems without significant resources. This is primarily due to the physical constraints imposed by existing technology and the designs embedded in commercially available CPU hardware. Essentially since its inception, the CPU has been constrained by the von Neumann architecture. After reviewing the silicon implementation of many AI paradigms, it is clear that hardware solutions can be configured to create scalable, efficient and reliable designs that can be re-configured dynamically and run as co-processor designs. Alternatives designs include using one of more core of a single or multi-core CPU, on the same motherboard or distributed across many systems.

## 6.2 Design Patterns

According to Fowler many people see *A Pattern Language* by Christopher Alexander's [2] as an 'important influence in the design world'. He wrote his patterns book

J.W. Tweedale & L.C. Jain: Embedded Automation in Human-Agent Environment, ALO 10, pp. 73–86.
springerlink.com                                                    © Springer-Verlag Berlin Heidelberg 2011

in a particular form which is known in the software patterns world as Alexandrian form. You can find good examples of this form in Domain-Driven Design, like Josh Kerievsky's *Pools of Insight* [190]. He continues with a description about the Gang of Four (GoF) [121], Portland, Coplien, *POSA* and *P of EAA* forms of Pattern Designs/ers [112]. Patterns enable programers to discuss design within the same lexicon, creating more concise understanding and more efficient use of resources.

Gamma et al. introduced patterns into the software domain in 1995[1]. This landmark work, often referred to as the GoF book, provided a list of 23 specific solutions, classified in-line with three forms of common design problems [8, 121, 387]. Developing multi-threaded applications requires the use of semaphores to reduce *hard-to-find* errors. Likewise, developing such applications with reusable components can be time-consuming to ensure efficient locking strategies are correctly employed. Strategised Locking and Thread-safe Decorator help developers avoid common problems when programming multi-threaded components or applications [327]. In 1967 Strachey [351] first described as *ad-hoc* or *parametric* polymorphism to describe the treatment of specific data type transparently, while in 1985 Cardelli and Wegner [49] modified this definition to include sub-classed genetic programming (supported by a single abstract super class).

When decisions are required to be made within an environment, a team would be assembled with the current resources and skills considered necessary to achieve the goal. If either resources or skills are lacking within the environment, the necessary skills or resources may be obtained from another environment or source. Once a team has been instantiated, agents within the team at each level are optionally able to collaborate with other agents in the team at the same level or with agents at a similar level in another team. Ordinarily, IA are constructed as individual threads and controlled using a team hierarchy, via commands being issued from a controller agents, at a level above, or delegated to subordinate agents at a level below, as illustrated in Figure 6.1.

## 6.3 Threads

Much has been written about threads over the past decades [223], however as multi-processor silicon becomes available to the desktop, the field has grown in popularity. Carver revives the theories of Tai [315] and has coded examples to support his arguments. The concept of multi-threaded applications requires a library of support functions to assist in the control and successful execution of user provided applications/threads. Portable Operating System Interface (POSIX) has become a unified standard (IEEE 1003.1) that provides this library which could be used to reduce the differences between multi-threading capable operating systems. The concept of Multi-Threaded (*OS/2, NT, Solaris and Unix*), Multi-Processor (*SPARC*), Symmetric Multi-Processor (*CRAY*) and distributed operating systems (*CORBA and Spring*) will also be discussed.

---

[1] Eric Gamma et al. leapt onto the software world stage in 1995 as co-author of their best-selling book *Design Patterns: Elements of Reusable Object-Oriented Software*.

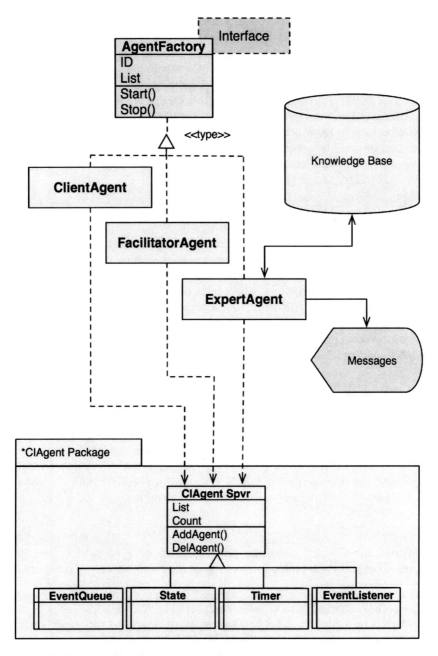

**Fig. 6.1** UML Representation of Demonstrator

Threads provide programmers with a powerful tool to enhance the interactivity of a graphical environment. Java provides a stack-based[2] Thread class that includes methods to start, stop, run and check a threads status. These threads may be synchronised using a paradigm based on a set of sophisticated synchronisation primitives, postulated by Hoare [155] around two decades ago, implemented on the Xerox PARCs Cedar/Mesa system. Java provides threads that are pre-emptive (based on time-slices) and includes a **yield**() command to enable threads with higher priorities to yield control to lower priority threads. Classes may be declared using the synchronized keyword, however they do not run concurrently. They are controlled using re-entrant monitor variables to ensure that all object variables remain in a consistent state when being switched [135].

## 6.4  Multi-threaded Applications

Lewis and Berg [420] provide a detailed discussion on Multi-threading in relation to *Solaris, OS/2, NT* and *POSIX*. They cover synchronization and locks, including: *"mutexes, conditional variables, read/write locks and semaphore"*. This discussion is augmented with a description of *"spin-locks, dead-locks, races, priority inversion and reentrancy"*. The POSIX IEEE 1003.1 Standard is also introduced, presenting a question on the effectiveness of monolithic operating systems designed using traditional top-down programming techniques. Especially when a collection of multi-threaded components that have dynamic capability could prove more useful.

## 6.5  Example

One of the early examples of programming *bifurcated trees* and *recursive algorithms* became classic examples on how to calculate/display fractals; such as those based on the *Julia* or *Mandlebrot sets*[3]. It should be noted that the same patterns can seen within a *Mandlebrot Set* after generating a *Julia Set*. Each series generally uses a variety of origins and magnification factors. Equation 6.1 states the basic function of a *Julia set* while Figure 6.2 depicts the resultant Julia fractal algorithm.

$$f(x) = x^2 + c \tag{6.1}$$

Similarly, Equation 6.2 states the basic function of a *Mandlebrot sets* and Figure 6.3 depicts the resulting cardioid fractal algorithm that can be generated on

---

[2] In Java a stack frame consists of three (possibly empty) sets of data; the local variables (for method calls), its execution environment, and its operand stack. The size of the first two stacks are fixed at the start of a method call, where the operand stack varies in size as byte-code is executed within the method [221].

[3] **Choas Theory.** The twentieth century was most famous for relativity, quantum mechanics, and the chaos theory. Fractals (Koch curves) are commonly used to display the turbulence catalogued using this theory. If a continuum, using a scale of magnification (from microscopic to macroscopic) is adjusted along that domain, it shows the various states of chaos [131]. The maths behind fractals is basic with a series of iterative calls that cause the curve to diverge if modulus $<1$ and explode if $>1$.

**Fig. 6.2** Julia Set - Parameters known as Dragon 9B [131]

any computer. In both cases, these fractals present repetitive patterns at various levels of magnification. Those shown in the resultant Dragon head and cardioid patterns clearly displays this repetition.

$$z_n = \sqrt{(z_{n-1})^2 + c} \tag{6.2}$$

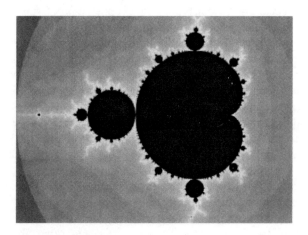

**Fig. 6.3** Cardioid Mandlebrot Set [131]

## 6.6 Component Technology

The concept of componentised software was first touted by M. McIlroy in 1968 during a NATO conference on Software Engineering in Brussels [249] . Szyperski [358] introduced a component as being an *object written to a specification*. It must exist autonomously and adhere to a published Interface Description Language (IDL). He and Messerschmidtt [252] agreed that for a component to produce a *predefined action or event*, the component requires at least five characteristics. These included: re-use, context, composition, encapsulation and serialization (versioning). The underlying communications model used can rely on COM, .Net, Corba's (ORB) or Java's (RMI). Examples include GUI Buttons, Java or Third-pary Beans or other self contained graphical objects used in an event driven enterprise application. The terms *wiring functionality* and *reuse* are commonly associated with this form of technology.

## 6.7    The Agent Transportation Layer Adaptation System

This communication component has been labelled the ATLAS model. It is based on the *Basic OSI Model* [169]. Significant effort on this topic has resulted in research into the specification of many syntactical (agent communication languages) and semantic (knowledge interchange formats) architectures in the quest of enhancing automated collaboration between agents in a distributed environment. Although this research has identified the requirement to develop alternative architectures for hardware, software and communication protocols. Nevertheless a number of concepts have been examined to determine the viability of continued effort in developing the suggested concepts. Most languages used to co-ordinate and collaborate were proven in a BDI style architecture using a team hierarchy based on the TNC framework, which is discuss in section 6.8.

Agents can be designed as monolithic solutions capable of autonomous processing of specific problems. These designs have very limited scope to solve problems other than those specified without significant programming effort. A team of less capable agents can be designed to achieve a wider range of processing, however their hierarchy is fixed and the system needs programming effort to change its functionality. Both designs were used in a controlled environment to assess the way agents can and should communicate. A series of autonomous agents classes were programmed at design time that could be instantiated at run-time by an operator. Each would communicate via a path (or queue)[4], until encountering a controller agent[5]. The controlling agent possessed the language capability of every agent that could be instantiated. Initially each language protocol used a specified port to indicate its type[6]. Each agent interaction was associated with a predefined task, which parsed the solution back to the originator when solved by the controller. After proving the concept of mixed functionality, effort moved on to improve trust and negotiation.

Trust may be obtained from another environment or source. Hence, using the closed-world-assumption that the original model or system has little or no prior knowledge of any new environment(s) it is proposed that agents be used to cross these system boundaries and to establish new trust relationships. Trust may be obtained from another environment or source. Hence, using the closed-world assumption that the original model or system has little or no prior knowledge of any new environment(s). It is proposed that agents be used to cross these system boundaries and to establish new trust relationships [374]. Agents may use different communication languages or knowledge interchange formats, ontologies and semantics to effectively communicate with the new environment. The trust model comes into play in order to establish a relationship, built on available information or adapted over time using an ongoing communication process that is compatible with that new environment. Such

---

[4] With the agent re-entering the environment from a point relative to that from which it paused or was suspended.

[5] For this model it is labelled the *babel fish* after the universal translator in the *Hitch Hikers Guide to the Galaxy* TV series.

[6] The development of a super translator to achieve this task for a predefined number of languages is still envisaged.

adaptation requires time for both systems to reflection and evaluation of the consequences of any past communication performed while operating in that environment (possibly by introducing a reinforcement learning system). Due to existing research in this area, effort quickly shifted to improving the process of negotiation between agents to improve context history and the trust relation.

As there are many ways of getting information into and out of a computer (*hardware*)[7], being a distributed application, this design is based on polling approach. The model continuously polls the environment for any form of communication and attempt to auto-negotiate meaningful communications to successfully conduct any transactions required to achieve the goals of the entire system. The flow chart of the *port scanning* process is shown in Figure 6.4.

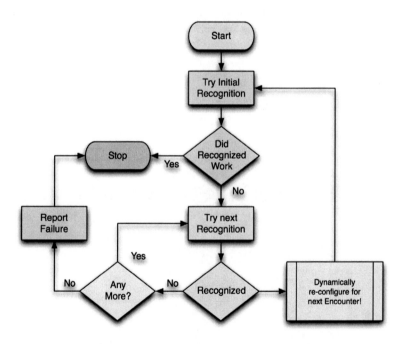

**Fig. 6.4** Polling Process [373]

Interchanging messages between different agent systems situated in different environments possibly affects all layers of the communication model, including transport protocol/system ports, ACL, semantic layers and Ontologies. To establish trust relationships with alien agent instances, the systems architecture must provide sufficient means to each agent instance, in order for agents to be *multi-lingual*. Without a multilingual capability, a dynamic agent team would be constrained and may compromise the goals of this text.

---

[7] These include memory addressing, port addressing, via the system bus (*via peripheral slots*), using either polling or interrupts.

Establishing communication between heterogeneous agent systems is crucial for building trust relationships between agent instances in these systems. Figure 6.5 illustrates the operator control panel generated by the ATLAS software. Incoming communication requests can be forked to autonomous threads capable of handling specific protocols for further processing and translation. To build a *proof of concept* for the communication layer of the TNC model we implemented a multi threaded client/server system to simulate the processing of multiple signal sources. These input stimuli are described by clients connecting to a server using a random protocol in our function model. The server thread models the communication interface of the receiving agent. The servers master thread accepts connection which forks as required, creating new threads, specifically to support each new protocol uncounted. The received information is processed and committed to the knowledge base identified during the *join* process. The communication prototype is written in Java using Sun's TM system library for multi-threading and networking. The function model is a simple Java application implementing the above architecture and visualising the on-going communication.

**Fig. 6.5** The Prototype Port Scanner GUI [373]

## 6.8   The Trust, Negotiation, Communication Model

Imagine an egg totally immersed in a pool of water, acting as a medium, in which messages pass. The thin shell of the egg dynamically forms flexible and robust interfaces that are connected to other IDLs to form a communication channel upon which data can be streamed. The white of the egg enables the entity/system controller to autonomously negotiate the type, form and mode of communications based on a context and level of trust imposed by the yolk of the egg.

The concept of the TNC model is shown in Figure 6.6, which details the type of interaction possible between single agents [280, 374, 408]. The model works for a hierarchy (team) of autonomic agents and by loose association for another agent team[8] within a complex system. The goal of the model being to provide a flexible structure that enables agents to team together without prior configuration (adaptation). Trust is the centre attribute used to form and maintain the partnership(s) and is required for agents joining or already within a team.

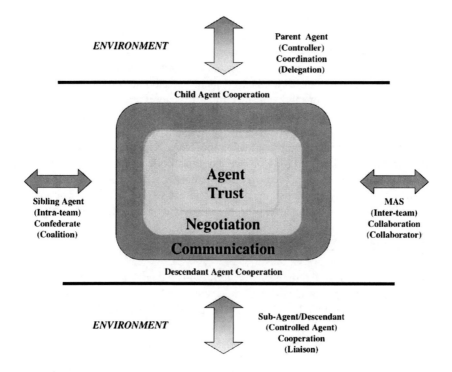

**Fig. 6.6** The Proposed TNC Agent Model [374]

The model is effectively a series of wrappers for agents that extend the communication ability that is inherently necessary for agents to incorporate an ability to *independently* negotiate. The model provides a communications interface, negotiation mechanism and a trust monitoring capacity. An environment would be built with a predefined list of resources and embody the skills necessary to create the required context at the time of instantiation. When a resource is called upon, by another agent or environment, a team (comprising of one or many agents in the form of the proposed model) would be instantiated by the resource manager. Each team would have a hierarchy, where each agent would fulfil a specific skill. The partnerships within

---

[8] Or human beings in future systems.

the team would be coordinated by a controller agent and the level of control based on loyalty through the bond established[9].

The approach is analogous to that used by humans. When Human decision makers are presented a problem, they are not able to solve themselves. They may choose to form a team of people they trust to assist in making that decision. The team is likely to have a hierarchical structure of some form in order to effectively manage the project with respect to scheduling or timing. The structure may include multiple levels. The TNC model is based on a similar concept.

When decisions are required to be made within an environment, a team would be assembled with the current resources and skills considered necessary to achieve the goal. If either resources or skills are lacking within the environment, the necessary skills or resources may be obtained from another environment or source.

Once a team has been instantiated, agents within the team at each level are optionally able to collaborate with other agents in the team at the same level or with agents at a similar level in another team. Ordinarily, agents are controlled using team hierarchy, with commands being issued from controller agents, at a level above, or delegated to agents at a level below.

In previous research on communication the Trust, Negotiation, Communication (TNC) model was discussed [374]. Since that time, collaboration with Quteishat et al. [301] explored as series of measures and methods that could be used to cooperate with one or more unconstrained agents that attempt to join the team. The joining process is constrained by the level of trust extended to or earned by that entity.

## 6.9   Assessing TNC

One approach used for building trust, is to provide a confirmation/feedback mechanism. A description of a socio-cognitive, agent based, model of trust, using Fuzzy Cognitive Maps is provided by Castelfranchi et al. [51]. Trust, in any model, is based on different beliefs which are categorised as internal and external attributes. The TNC model is based on the premise that the origin and the justification of the strength of beliefs come from the sources of the beliefs. Four possible types of belief sources are considered: direct experience, categorisation, reasoning and reputation. The credibility value given to an important belief is shown to influence the resulting decision, and the trust worthiness of the system. In addition to these beliefs, loyalty and fragility will be measured, based on the priorities used to form the bond. Kelly [188] discusses this process using objective and subjective measures. The later can be measured if its origins can be rated. It is proposed that loyalty will be used to measure the strength of the partnership and determine the frequency and level of monitoring, required to maintain that partnership. A measure of the level of loyalty can also be used to weight the bond and merit of information exchanged. The TNC model illustrates the collaborative bond between siblings within a team of MAS. The supervisors would have

---

[9] The agents within the team would also have the ability to seek partnerships outside of the direct team when they require or are able to provide additional resources. The partnerships established would be based on trust, which would strengthen or wain over time.

a higher priority to that of another team member or the human-computer interface. The co-operative bond, however, is hierarchical and implicit within that structure[10].

No metrics have been identified to assess the TNC Model, especially those used to measure trust and negotiation. Existing MAS systems will be assessed prior to choosing the appropriate candidate to embody in a prototype. However the TNC as shown in Figure 6.6 depicts one approach that appears capable of providing such metrics. The existing metrics for such models will be used to make MOE and Measures Of Performance (MOP).

### 6.9.1 Agents Threads

Each agent is running on a thread and there are two types of events associated with each agent: synchronous events that occur every N amount of time and asynchronous events that occur on the basis of events or incidents that trigger those events. An example of asynchronous event is the messages that are transferred between agents.

### 6.9.2 Agents Messages

Messages are the way of communication between agents. The messages follow a similar format as KQML messages. The messages have primitives that define the actions or states of agents. Depending on the actions the content of the messages includes the task that they should perform. For example, a message has the primitive of translate and the content of message is the word that needs to be translated. Each message includes a sender and receiver that is the agent sending the message and receiving it respectively. A Java class is defined to hold these values.

### 6.9.3 Facilitator Agent

Facilitator agent is a unique agent that has the list of all the expert and client agents once they are created. The moment the program is started some expert agents are initiated and they would send messages to facilitator using the Advertise performative for their message in order to inform facilitator of their expertise. The facilitator agent would then add them and their expertise to the list of experts or community of experts in other words. Therefore, when the client agents approach the facilitator and ask it to recommend an expert for a particular expertise, the facilitator agent would then refer the corresponding list and find an available expert agent to refer to the client agent. The facilitator agent can also keep track of the state in which expert agents are and take that into consideration when guiding client agents. For example, facilitator agent would know the availability of expert agents and therefore introduce the free expert agents to client agents or queue the client agents for the next earliest available expert agent.

---

[10] This is analogous to load sharing verses task management/processing.

### 6.9.4   Scenario

The scenario depicted in Figure 6.7 represents an unknown environment presently hosting an number of autonomous agents. Three remote agents (R1, R2, and R3) interact with the Communication Interchange Agent (CIA). Each of these remote agents could have different functionality and would normally be allocated a task required to contribute towards achieving an overall goal. For example, some of the basic functionalities considered in this research include: navigate, avoid obstacles, and seek resources. Given that a CIA agent may only have some of the prescribed capabilities required to achieve a task, that agent will need to communicate with other CIA agents in order to succeed. The agents would be using a predefined communication language. Some standardised languages to be considered include: (KIF), KQML or even FIPA. This particular scenario considers the use of a Radar sensor to detect targets. Several unmanned vehicles trying to locate these targets. The Unmanned Vehicles can be of three types:

- Unmanned Aerial Vehicle (UAV)s or Unmanned Aerial System (UAS);
- Unmanned Ground Vehicle (UGV)s; and
- Underwater Unmanned Vehicle (UUV)s [235].

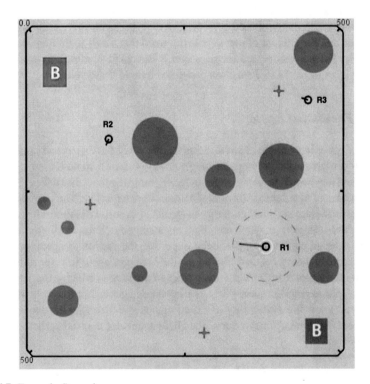

**Fig. 6.7** Example Scenario

A future case study will consider UAVs and UGVs in a hostile environment. Like the CIA example, these *Unmanned Vehicles* will traverse over specified areas to perform reconnaissance and identify targets. In order to achieve this role, they are required to communicate with each other to share their location, knowledge of the environment and more detailed information on each target. Information exchange could be constrained by the type of sensor or capability. However based on the combined knowledge-base of the Unmanned Vehicles in a specific area, the location of any target can be derived more accurately by the command and control station. Shared processing would require:

- multi-path information exchange (possibly using XML);
- knowledge-based discrimination and analysis; and
- complex data integration functionality.

Data streams from fewer specialised *Unmanned Vehicles* (fitted with cameras or Radar), can be used more effectively to target objectives within a defined geospatial region[11]. This information may also be analysed by a command and control station and enhanced with additional information streams to multiply the effectiveness of all resources deployed in that region or to assist in coordinating additional courses of action.

With the success of Project Nervana, the sub-tactical surveillance, intelligence gathering and communication capabilities of UAVs have been shown (Aerosonde). Clearly, using a number of UAS you can increase situation awareness. The enhanced reach provides a more immediate ability to detect objects of interest, improve decision making and response times. One or more UGVs integrated into such networks, can respond to objects detected within their proximity. To achieve this premise, It is important that all agents communicate effectively and share their knowledge effectively. Any number of UAS, UGV or UUV should be able to join or leave this network as the scenario requires (based on the resources available) and be capable of interoperable behaviour.

This scenario will be simulated using standard platforms and tools, such as virtual UAV/UGV in Simulink, in a synthetic environment before its physical implementation. The system will be tested for effective coordination, communication and knowledge-sharing between the different *Unmanned Vehicles* with autonomous behaviour. The communication between these Unmanned Vehicles will be implemented using an Agent Communication Language.

The examples discussed here illustrate the potential of each of the techniques examined. It is clear that learning, coordination and cooperation should also be featured in this text, however only a limited number of topics can be covered. Using a blackboard model, a number of techniques can be progressively added by researchers over time. This approach improves reuse and retains some ability to retain existing metrics and benchmarks. The enhancements provided by progressively integrating new models can be systematically applied to demonstrate the feasibility of generating autonomous teams in component form.

---

[11] Highly integrated, low cost, light weight, autonomous vehicles, would provide significant savings in logistics, manpower and resources.

*"No problem can be solved from the same level
of consciousness it was created [287]."*

`while`

*"Everything should be as simple as possible, but
NO simpler [264]!"*

Albert Einstein

# 7

# Improving Agent Architectures

Agent technologies have been an ongoing field of research that has failed to produce a solution for a complex, dynamic, real-world problem to the scale and expectations the technology is perceived to deliver. Web based-applications represent shop fronts and transactional services. However due to security, response times and reliability, industry has chosen to avoid them for mainstream advanced information process technologies. The KES centre has conducted reviews on agents and agent based systems and concluded that industry requires a BDI framework within an interoperable team-based environment. After reviewing projects using R-CAST and JACK like Integration of Reactive Behavior and Rational Planning (InteRRaP) [260], alGOl Logical prOGramming (Golog) [222] and RETSINA [145][1], the risk of mode confusion by human operators surfaced as a critical issue[2]. In many cases, load sharing can alleviate this problem, hence the requirement for agents interoperability, especially between supervisors, siblings and subordinates. Trust and negotiation become important factors in establishing context, cognition and task related activities. Such attributes are normally coded or hidden within an applications corporate processes. Autonomous teams with the ability to dynamically alter functionality present ideal enhancements to any scenario. Modularised capabilities must be, proven available before they are adopted by industry. to achieve reusable, hardened applications.

---

[1] This project is now incorporated into the CoABS grid to gathering information or constructing plans useful for decision making and goal success.

[2] Under normal time pressures, human teams normally make the correct decisions about a potential threat, but when they are subject to time constraints, the same teams' performance may suffer due to the lack of information sharing. Under stress conditions, the likelihood of an incorrect decision escalates.

J.W. Tweedale & L.C. Jain: Embedded Automation in Human-Agent Environment, ALO 10, pp. 87–104.
springerlink.com                                              © Springer-Verlag Berlin Heidelberg 2011

## 7.1   Background to AI Research

At present there has been a host of work surrounding trust, negotiation and communication. The research on trust has been either at a HMI level or to promote e-commerce. Examples of research into agent communication continues to attract discussion within the agent community. This started with ACL and concludes with SOAP as the most expedient protocol for Web based exchanges. The concepts evolved during the development of mainframe technology using the *skeleton* and *stub* approach to exchanging data, internally or externally has also been exploited by Corba (Orbs), Java (RMI) and languages supporting Web-services, (using what is now called .NET). The experiments conducted prove the need for more than one stream of communications. This enables both static and dynamic information updates. It is extremely important to revise the existing protocols to enable data, information and knowledge to flow freely, but more importantly, efficiency within the system. Research into this approach is still immature and must be the focus of future enhancements.

## 7.2   Existing AI Research

Data is a quantized collection of facts from sensors, devices or condition taken from within the problem related environment. It can be stored, observed, analyzed and/or scrutinized to produce information prior to being compared against some threshold upon which a decision is inferred. A number of resulting Course Of Action (COA) can be decided based on this output. For instance, a KBS uses a knowledge-base extract from one or more SME by a knowledge engineer and placed in a repository. The incoming information is dynamically tested against a knowledge-base by an inference engine (based on a specified problem-solving paradigm) could be used to automatically make informed decisions/judgements about the function or process being requested. A review of the major problem-solving paradigms that needs to be included, follows:

- Knowledge-Based Systems,
- Case-Based Systems,
- Reasoning Systems,
- Model-Based Systems,
- Bayesian Networks,
- Fuzzy Logic Systems,
- Pattern Matching,
- Neural Networks,
- Evolutionary Systems, and
- Hybrid Systems.

It is possible to construct context-switchable MAS test-bed platform and establish the communication between agents in a distributed environment. These agents would be dynamically scalable and allow entities to travel across the virtual environment, becoming available to all clients. Various AI techniques can be fused into the system by adding agents with different capabilities including: reasoning, planning and learning. This ingenuity required a host of historical accomplishments relating to both technology and technique.

## 7.3  Summary

During the Pascal evolution, code was compiled (or cross-compiled) as a binary file run natively. Every time software was compiled, the programmer required knowledge about the hardware and platform specific constraints. This was very restrictive and resource intensive. Emulators quickly evolved, enabling the interpretation of inter-mediate code (like P-Code and Byte-Code)[3]. As software became more accessible, the data being collected started to lose its relevance. Data Warehousing and knowl-edge management flourished. In 1974, Buchanan and Shortliffe developed MYCIN and Latter DENDRAL [41, 105, 336] Other advances introduced the concept of OOP, reuse and agency theory. Architectures, like Enterprise Management and Modelling Architecture (EMMA) emerged[4]. Applications started to combine human-computer decision making to conduct *General Problem Solving*, using sub-system with well maintained knowledge bases[5]. As knowledge-bases matured in complexity, accuracy and acceptance, new methods evolved to manage and to represent incoming infor-mation visually[6]. Research has continued and many new architectures continue to emerge.

## 7.4  Microprocessor Architectures

Computers evolved from the perceived needs of human kind to automate or expedite computational problem solving that initially included cryptography and extended into office automation. Many classifications of computing have emerged, starting

---

[3] This form of execution is also know as Just-in-Time (JIT) compilation or interpreted in-termediate language.

[4] EMMA is based on research conducted under DARPA contract #F30602-91-C-0016 and the Office of Naval Research (ONR) under contract #N00014-92-J-1298. Its architectural design has six layers, with each layer reliant on the previous layers functionality. These layers include: the network, data, information, organisational, coordination and market-ing layers. As described these layers concerntrate on the information management system concept and omit the more traditional International Standards Organisation (ISO) layer approach, although the concepts and relationships are visible.

[5] Corporate knowledge data-bases were generated using inputs supplied by SME's.

[6] Wang laboratories attempted to create enterprise level applications, however dropped most of their revolutionary efforts after a price war broke out among workstation makers. The 200 researchers committed to the graphical workstation project were redeployed and the technology sold off to xerox. Little has been published about the resulting operating system, but project sources say that it was the beginning of modern-day component communica-tions using distributed resources [399].

with analogue and digital computers, reaching to the classification and employment of the types of circuits used. In many cases software can be expressed in terms of hardware using tools like Very High Speed Integrated Circuit Hardware Description Language (VHDL). This technique was investigated to increase efficiency and throughput of many applications, especially those using CI functionality.

Over time a number of generation of computer architecture were introduced (see Table 7.1). Most extended designs that previously emerged, but for some reason failed to scale in performance as the demand for higher capacities grew. These designs include the: Von Neuman microprocessor, Harvard microprocessor, complex instruction set computing, reduced instruction set computing, very long word instruction set, input/Output controllers, memory management, parallel/vector microprocessor, co-processor feasibility. Obviously computers have advanced beyond the $7^{th}$ generation, however no one has categorized these advances. We suggest that massively parallel microprocessors (using the System on Chip (SoC) concept)[7] and Graphics Processing Units (GPUs)[8] would be two candidates for inclusions. Mobile and hand-held devices could also be included[9].

**Table 7.1** Computer Enhancements by Generation

| Generation | Period | Technology | Example |
|:---:|:---|:---|:---|
| $1^{st}$ | 1954-59 | Vacuum Tube | IBM 1401 |
| $2^{nd}$ | 1957-64 | Transistor Cct | IBM 360 |
| $3^{rd}$ | 1965-71 | Microchips | PDP-8 |
| $4^{th}$ | 1981+ | CPU (VLSI) | Desktops |
| $5^{th}$ | 1989+ | MPU | Cray |
| $6^{th}$ | 1998+ | Super Scaler | Multi-core/pipe |
| $7^{th}$ | 2010+ | System Processors | MPU + GPU + I/O |

## 7.5  Abstracting Software into Silicon

Silicon has evolved over the past 60 years, however designs that adapt or can be reconfigured dynamically have more recently raised debate. During the infancy of AI small analog FPGA were being used to improve performance bottlenecks. Evolutionary algorithms were used to abstract out the analogous nature of the embedded logic. The increasing modularity of the circuit designs facilitated the decomposition of these new structures. Mead's exploited the concept with the inception of the *silicon retina* based on a complex time-continuous network of recurrent neural nets. Another experiment developed dynamic noise filtering (as used in early bionic hearing devices) and a third that dynamically smoothed extraneous measurements to reside with the constraints of a fitness function [364].

---

[7] Such as the i7 and Phenom II.

[8] Such as the Tesla and Fermi chip sets, amoung others.

[9] Such as Apple's A4 and its replacements.

There is of course the abstraction of AI techniques that have already been implemented in FPGAs as co-processor alternatives to CPU design or even a separate processing cores within same silicon die of current microprocessor designs. A number of existing AI techniques were listed in Section 7.2.

In 1963, Maccel P. Schutzenberger documented the deterministic PDA machine which was used by Robert J. Every to produce a TM using a push down Last In, First Out (LIFO) stack, push and pop commands enable recursive algorithms to run on more efficient or previously hamstrung technologies.

## 7.6 AI in Field Programmable Grid or Gate Arrays

Application-Specific Integrated Circuits (ASICs) or their equivalent analog Programmable Array Logics (PALs) and FPGA have successfully been used to embed a 4-input logic controller with up to 12-inputs, to produce up to eight programmable singleton values. Each singleton can be driven by a bank of 64 rules that all operate independently. Each circuit fuzzifies the input, determines the membership function and defuzzifies the output into a crisp value. The DSS process is achieved using rules, weights and Membership Function Generators (MFGs), therefore the configuration can be changed dynamically [230]. Using a Centre-of-Gravity defuzzification approach the outputs of the trapazoidal transfer functions can be 'ORed' by hardwiring their products to aggregate transconductance of the circuit being evaluated. Fast digital fuzzy processors using 2,401 inputs to provide up to 128 outputs operating at 50 MHz are now common, with Digital Signal Processor (DSP) designs being evaluated. A number of examples follow to illustrate this concept.

### 7.6.1 Analog Fuzzy Logic Controller

The first fuzzy logic controller was reported at AT&T Bell Laboratories in 1986. This was based on Zadeh's experiments which were validated by Mamdani and later pioneering by Yamakawa [413]. These events were studied by many researchers and resulted in a series of very simple non-linear applications. New digital processors using lookup tables were being generated to represent membership functions. Inference engines became pseudo DSS with antecedent and consequent clauses used to generalize the reasoning methods. Four types of membership are provided. These include: S-shaped, Z-shaped, triangular shaped and a combination of the three 'terms' in the trapezoidal function. Each has a series of different shapes and slopes. Max and Min circuits are also provided, resulting in a five 'term' fuzzification system. The resulting formation represents an autonomous mobile system that is capable of navigating the unstructured environment within the real world [143].

### 7.6.2 A Fuzzy Logic Control Taxonomy

The question of what can be gained by implementing a Fuzzy Logic Control (FLC) is best described by examining its taxonomy. Generally six benefits are discussed. These include the: Capture Operator or Artisan Knowledge (COAK), Exception

Handling (EH), Generalized Damping (GD), Local Adaptation (LA), Generalized Constraint Enforcement (GCE), and Input-Output Maps using Interpolation (IOI) [363]. This taxonomy is neither complete nor exclusive and will grow as the science evolves. No single tool or technique fits every situations, but the hysteresis enabled using fuzzy logic can be used to provide general feedback as one approach to facilitating *fine* and *coarse* controls simultaneously (examples include heating, cooling and other stepped control applications). The control policy or application must therefore be considered as the primary objectives in describing the FLC taxonomy.

### 7.6.3 A Fast Digital Fuzzy Processor

FLCs are commonly found in washing machines, cruise control and anti-braking systems because of their reliable output, ease of implementation and low cost. As technology advances, the response times also improve. One domain that has denied the use of FLC is that of controlling High-Energy Physics Experiments (HEPE) (time being a stringent constraint). Research into the current FLC architecture with the goal of implementing low cost (time) identification of rules through parallel-pipelining is being pursued. One technique called *Active Rule Selection* has been tested. Using 12 pipelines, throughput of 20 nanoseconds have been achieved [120]. By allocating fuzzy memberships as a 7-bit word, sufficient performance can be achieved to control each HEPE. Further precision will come with the speed and density of more modern VLSI circuits.

### 7.6.4 Bayesian Networks

A Bayesian Networks can be used in conjunction with 'graph' theory to determine the level of risk of using a computer for decision making. Automation is increasing using soft computing techniques to consider present situation awareness in a $C^2$ environment. Bayes' Nets adopt an Object Oriented Programming Software (OOPS) approach to solving real-world problems [9]. The technique employed recently includes three layers:

- objects of the scenario,
- the assumed importance of the connection, and
- the template that dynamically aligns the results with the topology [347].

Based on the network generated, a dependency graph is extrapolated to ascertain the essential connectivity required by the risk assessment tools (in this case an expert system).

### 7.6.5 Case-Based Reasoning System

Case-based reasoning has successfully been embedded in an FPGA using *Genetic Algorithms (GAs)* to recycle existing solutions into to solve new problems. The solution needs to initially ignore any constraints to obtain a robust solution. Constrained problems need to deal with the interconnection topologies of the FPGA hardware selected. The use of GAs is extremely computationally intensive, which constrains the ability to scale and increases the risk of producing errors (introducing the need for manual correction). The concept of selection, retrieval and adaptions with programmable DSPs makes processing analogue signals a viable solution for this technique [178]. The method of using configured functional cells can be catalogued and targeted to make this technique more robust and less error prone.

### 7.6.6 VLSI Knowledge-Based System

Modern day VLSI chips are constructed using clusters of cells, which contain functional architectures interconnected by fusible interconnects. A number of tools have also developed over time making VHDL design more efficient and reliable. Existing design environments like PALLADIO and CMUDA required catalogues of FPGA cells in database to build objects in *frames* [111]. Semantic links were used to develop *slots* that are resolved against a number of heuristic rules for any given design. Alternative description for each cell or cluster functionality, together with design libraries are used today. Although designs can be simulated using products like CRYSTAL or SPICE, a top-down floor-planning approach was developed using Prolog which aided in the reduction of path delays, circuit real-estate and power consumption.

### 7.6.7 O(N) Net Complexity

A successive set of non-linear, zero-level *winner-take-all* neural nets can be connected to effect the voltage across a single wire $O(n)$ long. This circuit behaves like an analog device without *hysteresis*, although require sufficient time to encode the size of its largest input value without ringing. By combining a number of circuits ($k$ Position), a complex computation can be configured based on a single input. CMOS VLSI capable of auditory localization with up to 170 inputs have been manufactured to measure the spatial order of each input [217].

## 7.7 Hardware Directions

As discussed, add-on hardware is advancing, making it possible to embed sophisticated CI designs into silicon and in many cases achieve significant performance gains over present software solutions. FPGAs can be used to host one or more designs and are extremely useful when implementing massively parallel processing like neural nets.

More modern silicon is based on flash memory technology to enable them to be dynamically re-configured at speeds that are approaching real-time. The latest generation of FPGA are designed as a complete SoC and can be used as a co-processor to any modern CPU. As described in the previous chapter, existing microprocessors are not designed to efficiently execute CI applications, and their manufacturers are enhancing designs for web and multi-media based applications. FPGAs are one solution available now, that is accessible to provide efficient, scalable, robust CI designs.

Similar enhancements are required to provide effective autonomy to support the interoperability demanded of systems currently being developed.

## 7.8   Open Architecture

Many systems fielded in the military have designs (using closed architectures) that are over a decade old before being commissioned into service. They use proprietary software that is often tied into the existing hardware and rely heavily on technical refreshers to implement new capability (technical insertions). Five challenges have been identified to address the move towards open business and architectural models. The first is to clearly *delineate* the software from the hardware. Next is to force project managers to *use the APB* methodology, and avoid obsolescence by planning for *technical insertions* based on Moore's law [257]. The other two challenges call on the premise *"design once, uses many time"* and *linking* the project to an operational or capability map [350].

## 7.9   Structuring Agents for Adaptation

Adaption is referred to a structural change required to process knowledge from a variety of sources to a single output via a common interface. ZEUS [282] and PARADE [22] are examples of adapting agent functionality using automated tools. Changes in context requires new communication languages, protocols [384] or learning capabilities. Splunter uses an agent factory to instantiate BDI agents [33, 128] using a coordination pattern design within a *black box* model with a predefined number of slots that accepts component that can only be inserted into *agent-specific-task-slots* designed to operate with the web-services architecture[10]. Examples include QUASAR [19][11] and KIDS [342][12]. This calls for an analyst to decompose the task or goal and pre-allocate resources as required[13]. Team coordination is achieved using cooperation and collaboration, but each function is achieved at different levels of information. This is also true using BDI which is targeted at the coordination level, where information flows are derived from cognitive knowledge. As tasks are allocated using patterns that can be encapsulated as classes running on one or more threads, the teams functionality could be configured as a composite component, however it needs to communicate across several levels using a number of protocols simultaneously.

---

[10] Using RMI, WSDL and DAML-S.

[11] The QuASAR project aims to provide a toolkit to assist in the cost-effective creation and evolution of reliable semantic annotations Web services.

[12] This system provides tools for performing deductive inference algorithm design expression simplification, finite differencing, partial evaluation, data type refinement, and other transformations.

[13] This functionality can be switched using a number of methodologies, from *IF-THEN-ELSE* blocks, *CASE or SELECT* statements and more recently using *polymorphism* based on types.

## 7.10 Dynamic Programming

Mathematicians once solved problems by systematically finding one result after another. As early as 1982, Kohler used software to evolve the best way of wining *darts*, but it consumes a serious number of resources and processing power. Based on problems representing exponential solutions, hardware algorithms can be used more efficiently. Subrata [74] describes the ability to store and recall a large number of emulators (programmed in firmware as micro-codes in a CPUs micro-rom) into a local target mechanism is generally called *dynamic micro-programming* [141]. This concept is also referred to as the ability to create disparate *exo-architectures* [261]. Modern microprocessors exhibit new internal functionality with each revision. Examples include multimedia and signal processing extensions. The new Intel Core CPUs now exhibit re-writeable micro-code.

## 7.11 AIP Technology

Very few systems provide complete solutions and for this reason generations of development occur. One goal of re-use is for each new generation to extend rather than replace existing functionality. New technology enables alternative techniques to be developed and it becomes a matter of time before these additions are integrated[14]. This domain grew to significance, although the author of the terminology has since admitted that he would have chosen the term CI to reflect its true capacity. AI is based predominantly on OOPLs. Confusion surfaces when designers use UML descriptions, such as; *aggregation* and *composition* when decomposing problems. Abstraction enables the programmer to aggregate classes[15] which can be composed[16], were inheritance *extends is-a* as a specialized part of object and an *interface* makes that component which *look-like* something else.

As discussed, the design of OOPS uses an iterative process based on a strong system engineering methodology. The design of AIP technology uses a structured framework. The fundamental concepts include:

Performance:        AIP technologies are generally capable of solving many problems quicker than the time it appears to press a button. When humans are included in the process, the performance and interaction is based on response times provided or accepted by the operator. This form of functionality no-longer relies on the number of instructions the system can process per second. Alternatively, some system based stimuli are time dependant. As time dependent applications need a response within a specified threshold, agent based decision making becomes a viable alternative source of response or clarity.

---

[14] AI was born from within the field of mathematics and was manifested using software.

[15] WClasses which associate whole-things, that *uses-a* component or data type.

[16] A composition is represented as a *has-a* relationship where the object is part of a larger object.

Reliability:       The assistant shall have built-in hardware and software ele-
                   ments that are designed to reduce the risk of a complete sys-
                   tem failure. The applied technologies should allow for grace-
                   ful performance degradation in case of failure.

Modularity:        The assistant shall be based on technologies that allow log-
                   ical decomposition of the system into smaller components
                   (modules) with well-defined interfaces. Modularity facili-
                   tates development, enables future upgrades and reduces life-
                   cycle costs by improved maintenance.

Integration:       The assistant includes many diverse functions needing dif-
                   ferent implementation methods and techniques. The technol-
                   ogy used should support integration with conventional, as
                   well as advanced, methodologies preserving modularity.

Maturity:          The assistant shall be based on mature and proven implemen-
                   tation technologies. This is expressed by the availability of
                   tools, successful prototypes and operational applications.

## 7.12  Trust

Trust is encountered across a broad domain of applications and perspectives with
levels of inconsistent influence. A unified definition of trust is elusive as it is context
specific. However a number of notions appear commonly in the literature on trust including: complexity [65], reliability [300] flexibility, predictability [125] credibility
[115], complacency [188], consistency, situational, experiential, security, accuracy
[115], dependence, responsibility and uncertainty. The notions cover a spectrum from
the system (local) level. For example, the usability or reliability of a specific piece
of software, through to a grand social level, such as responsibility and dependence
[156]. In the following sections some of these issues are discussed and their relevance
to the current work considered.

In Kelly et al. [188] trust is defined simply as the confidence placed in a person
or thing, or more precisely, the degree of belief in the strength, ability, truth or reliability of a person or thing. They also list a number of elements of trust identified
from the research literature, including faith, robustness, familiarity, understandabil-
ity, usefulness, self-confidence, reputation and explication of intention. As is evident,
the dimensions of trust are many, and it will be important to narrow them down for
the current context.

Perhaps the most important factor in trust is risk. The ability to reason about trust
and risk allows humans to interact even in situations where they may only have partial information [51]. If the risk is believed to be too great the interaction may not
take place. Three distinct components or levels of trust are presented in which at each
level there is an increasing tendency to believe that a person is trustworthy. The levels are: predictability (Used in the early stages as a basis for trust), dependability
(Corresponds to the trust placed in the qualities attributed to the other entity), and
faith (Reflects an emotional security). The dynamic nature of trust is self-preserving

and self-amplifying [51]. It is increased through successful interactions and degraded through unsuccessful outcomes. Some level of trust will be gained if a system meets certain expectations which may be different between individuals, and the situation.

As mentioned, the trust layer within the TNC Model is the basis by which teams are structured and by which the partnerships between agents are formed and maintained. Using trust is again analogous to human teams. Bonds or partnerships will be formed based on the trust that members have sufficient capabilities to solve the problem, will provide a solution within a sufficient amount of time, and so on. Trust is required to initiate a partnership, to remain within a partnership and to resume/re-initiated a partnership. Further detail of this trust layer will be explored as the model matures.

### 7.12.1   Human-Computer Trust

Trust is increasingly becoming a factor in the relationship between humans and machines, and in particular, automated systems. Although trust between a human operator and an automated system can be significantly improved using Human-Computer Trust (HCT), is not the major focus of this chapter. The TNC model proposes to address how trust bonds can be established, monitored and shared [374]. Schneidermans experiments confirm that the trust between entities is separated by the expectations that entities have about a process or object and therefore different approaches need to be considered when dealing with trust issues within a system [334].

HCT is one of the limiting factors in delegating more responsibility to automated processes[17]. This concept has been documented in procurement [353], computing [156], on-line interactions and e-commerce [318], transportation [328] and security domain [45]. That bond may be perceived by some[18] to be trustworthy due to the perception that they are autonomous and non-judgmental [156]. A fitting definition of human-computer trust was reported by Kelly [188] based on that originally described by Madson [232]. He states that trust is the extent to which an entity is confident in, and willing to act on the basis of, the recommendations, actions, and decisions of other entities generating decisions. Muir crossed the taxonomies of Barber and Rempel to form a single two-dimensional taxonomy of human-computer trust. This shows that trust is an important intervening variable between entities involved in automated processes [192].

The ability to act without human intervention, autonomy or other systems, is a key feature of an agent. Despite the lack of a universal definition, we can utilize the properties that an agent is supposed to have. This could help us classify specific its type. Franklin listed some properties of an agent shown in Table 7.2.

---

[17] The human agent is not able to provide a consistent focus on promoting loyalty and the trust bond suffers.

[18] Usually beginners who lack the skills to challenge the results or posses the capability of guessing an outcome based on previous experiences.

**Table 7.2** Agent Properties as defined by Franklin [116]

| Property | Meaning |
|---|---|
| reactivity | responds in a timely fashion to changes in the environment |
| autonomy | exercises control over its own actions |
| task-oriented | does not simply act in response to the environment |
| sustained | is a continuously running process (temporarily) |
| communicative | communicates with other agents, perhaps including people |
| adaptive (learning) | changes its behavior based on its previous experience |
| mobile | able to transport itself from one machine to another |
| flexible | actions are not scripted |
| character | believable personality and emotional state |

### 7.12.2  Trust in Decision Support

Machines may be optimised to make sound decisions; however, operators don't necessarily believe what they're being told. Some level of trust is required between all entities, be that human-to-human, human-to-machine, machine-to-machine or machine-to-human, before any decision can be successfully communicated. The decision must be communicated with clarity and accuracy, using appropriate and timely feedback. The interface must be intuitive enough to show sensitivity or retain focused attention on the goal or decomposed task(s). The interface must also cater for ad-hoc relationships. This would personify the processes that humans use to form friendships (usually by establishing mutual bonds). Alternatively it is seen as encoding the metaphor of man-machine communications/trust. Suchman [354] describes the lack of mutual understanding as the fundamental cause of miscommunication errors, such as *false alarm* and *the garden path*. He concludes by describing an effective HCI, as one that must include an understanding of situational context relating to real world knowledge with broadband communications (visual, verbal, and non-verbal). Moray [258] on the other hand suggests designers need to discover the users functional mental model and provide information as and when expected. Chambers and Nagel [53] also completed a study based on automated pilots and concluded that the human/controller is ultimately responsible and therefore cannot be left out-of-the-loop. Brown et al. [37] concludes that to provide optimum performance of any system, it is necessary to maintain all entity relationships. This can be achieved by concentrating on issues such as: authority, trust, functional allocation and automation.

### 7.12.3  Planner/Scheduler

The planner/schedualer in the National Aeronautics and Space Administration (NASA) DS1 spacecraft is constraint-based and uses an integrated temporal planner with an embedded resource scheduler. The planner uses a heuristic guided backtracking search algorithm to produce flexible temporal plans in real-time. An example of this concept on DS1 is the simultaneous maintenance of navigation, image acquisition and gimbal control (for solar panel attitude).

This merger of deduction and reactivity has been questioned due to the ability to bridge the temporal requirements. By minimizing inference[19] doubt may also be cast upon the validity of using *deduction and search*[20] in such models. Especially with a layered model of time-limited mini-plan libraries that represent each segment of the causal model. The resulting system must therefore be a single model that is reactive, concurrent, simple and deductive with *embedded truth maintenance*.

Other lessons learned include the need to retain a capability to include the human-in-the-loop, the validity of test and validation prior to launch, schedule impact of automated systems over the traditional flight schedule and the endeavor to support collaborative synthesized autonomous systems in future research domains.

## 7.13 Negotiation

In the true sense there are two major forms of negotiation coded in applications. These include various forms of transactional and dutch auctions. Trust is a major component of agent based negotiation, however in most instances it is used to decipher the amount to be bid, how many increments and other limits that industry can use to maximize the final transaction value. A good example is the automated bidder on amazon. All of these forms of bidding are open to external influence, especially mechanisms used to control or maintain the bidding price in favor of the seller. Although this form of negotiation is valid, it may also be used to automate the process of self-discovery and used to arbitrate the language, protocol and ontology required for efficient communication. Most of these attributes are assessed during design, coded at design-time with problems generally being manifested at run-time. In an ideal world, it would be convenient if everyone were to communicate using the same language, semantics, protocols and taxonomy. Alternatively everyone would be capable of translating these aspects of communications in real-time. As the babel fish is fictitious, some other form of translation must be available when negotiation communications needs across a standard interface. This interface is not the topic of this text and assumed to be the same for all connections unless otherwise stated.

## 7.14 Communication

The evolution of agent based communication has been discussed in Chapter 5. The concept of using communications as a key aspect of the TNC framework relies on the ability to discover, self-synchronize and autonomously establish a working level of trust in a given environment. Technologies are already available to support numerous protocols, a messages syntax and semantics, as well as a variety of transmission modes or mediums. This discussion is confined to agent messaging systems and supporting architectures.

---

[19] Using combinatorial circuits in a subsumption architecture.
[20] Possible through the use of *conflict-directed best-first search*.

## 7.15   Component Messaging

Specialized techniques have been developed to enable the communication of agent based software components and the exchange of information between specific applications. Agents can communicate by sending messages directly to one and another using a prescribed format with rigid semantics. These messages are accepted by all the agents within the MAS and actioned as required [385]. Standards are required in order to establish compatible communication among any agent that gains access to that distributed MAS environment. Several common standards include: ACL, KQML, FIPA and SOAP. Communication includes the delivery of complex attitudes or behaviours [209] through a specified architecture [375]. Therefore there is a requirement to identify the: syntax, semantics and pragmatism used, especially the symbols. Agents are able to communicate with other agents, using a specified precedence and hierarchy [90], however the need for immediate and broadcast messaging has becoming common place.

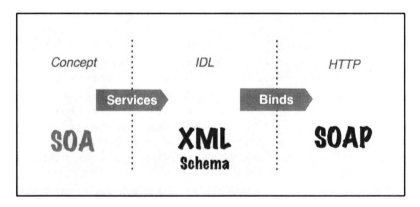

**Fig. 7.1** The SOA Process

## 7.16   Distributed Architecture

It is important for agents to gather knowledge about its environment to enable it to make intelligent decisions and actions [25]. Agents can be equipped with sensors to collect information. Communication between agents can be conducted directly between two agents or through a facilitator or interpreter. The exchange of information in a particular domain of knowledge that requires each agent to have shared knowledge of concepts in a particular domain is known as ontology. Each domain of knowledge would have its own ontology, although this can become flexible using more modern protocols. SOAP is an inter-application cross-platform communication mechanism. It defines a simple and flexible communication format that is based on XML documents passed across HTTP [331]. XML is used to represent data as an object instead of symbols and is passed in a standardized text-based format [21].

This process is depicted in Figure 7.1. Using the SOAP protocol, information is packaged in SOAP messages that consists an envelope and body, with an optional header elements [136].

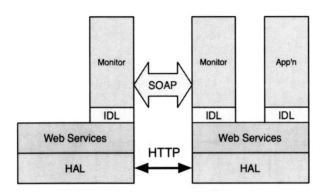

**Fig. 7.2**  Application Based on SOA Structure

SOA models are constructed using loosely coupled components based on Web Services (Beans, Components or Enterprise Applications). Java classes are deployed as Web Services to consume targeted objects. Monson-Haefel (2003) postulated that Web Services, like distributed computing, is: "simply the hardest problem in computer science [146]". SOA provides an abstracted service at any granularity, which is great for hiding an entire system or sub-system. It can also become hierarchically dynamic using web service. Figure 7.2 displays the physical architecture invoked in each computer to achieve distributed messaging based on the SOAP framework. SOAP should have been a simple answer to interoperability, however the lack of proper standards initially made the job of distributing objects more frustrating than necessary. XML, SOAP, WSDL and Web Services retain the key topics in making successful distributed applications, although seamless integration of JAX-WS and JAXB is required to achieve the low-level functions and interfaces.

The SOA capability was included in Java 5.0 (Mar 2006) and was updated in Java 6.0 (Dec 2006). These Application Program Interfaces (APIs) reduced the expertise required by programmers to use Java Web Services in their SOA applications (Knowledge about *Generics* and *Annotations* would aid productivity). Although developers will need to update their knowledge using SOA, cross compilation of WSDL to Java and deployment before launching into creating a distributed application. Remote procedure code was included in Java 4.0 (2002) with Web Services, Binding, Meta-data and Run-time behavior undergoing significant revision to trigger the *tipping point* of acceptance and wider spread implication.

## 7.17  The Dynamic Architecture

Establishing communication between heterogeneous agent systems is crucial for building trust relationships between agent instances in these systems. Interchanging messages between different agent systems situated in different environments possibly affects all layers of the communication model, including transport protocol/system ports, ACL, semantic layers and Ontologies. To establish trust relationships with alien agent instances, the systems architecture must provide sufficient means to each agent instance, in order for agents to be *multilingual* as shown in Figure 7.3.

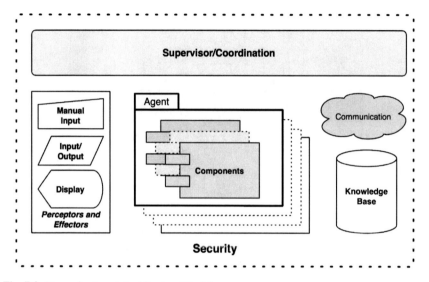

**Fig. 7.3** Dynamic Agent Architecture Model

Incoming communication requests can be forked to autonomous threads capable of handling specific protocols and further processing and translation. To build a *proof of concept* for the communication layer of the TNC model, we implemented a multi threaded client/server system to simulate the processing of multiple signal sources. These input stimuli are described by clients connecting to a server using a random protocol in our functional model. The server thread models the communication interface of the receiving agent. The servers master thread accepts connection and forks it into a new thread specific for the determined protocol type, the received information processed and committed to the knowledge base. The communication prototype is written in Java, using Sun's™ system library, for multi-threading and networking. The function model is a simple Java application implementing the above architecture and visualizing the on-going communication.

## 7.18  Micro-simulated Capability

People rely on automated systems to make decisions, especially when time lines or expectations are compressed. Three levels of confidence and embedded expertise is being used within the system to affect any response used to inference or control the environment. Humans prefer simplicity and they attempt to streamline most processes to avoid stress, This may result in them skipping the "cognitive effort required

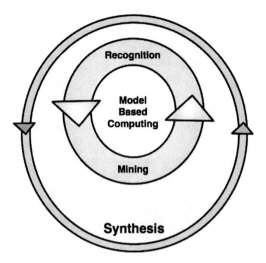

**Fig. 7.4** The Concept of Micro-Simulated Decision Support [378]

to gather and process information" [57], and rely on intuition [371]. This method of coping often leads to mistakes via mode confusion, especially during intense periods of mission critical operation. *Automation bias* best explains the misuse of automated decision aids, although it cannot account for the lack of use in cases where it may have assisted [91].

Better training techniques and alternate models are being researched to solve some of these issues. The use of simulated reality is not new, however the practice of rehearsing the outcome of possible courses of action to enhance the solution being sought. By analyzing the combined operators estimates and/or beliefs during subsequent scenario cycles, data can be analyzed in faster than real-time to extrapolate missing variables or segments of a specific situation to gain an advantage in a variety of situations. To achieve this milestone, a collection of rules, system state and algorithms must be available in order to simulate the reality being attempted.

Thus using a combination of data analysis, mining and synthesis techniques in micro-scenarios, researchers can employ an iterative approach to problem solving, using a variety of technologies or scenario instantiations that operate in parallel across a distributed environment to promote this enhanced concept of IDSS initially described as Micro-Synthesised Simulated or Micro-Simulated Decision Support [378]. However Intel has combined a number of processes it adopted called Model-Based Computing (MBC) as shown in Figure 7.4. The technique involves: the analysis of a volume of data to create a feasible model, use data to test instances of a parameter against a model and create an instance of a potential model. By increasing the number of processing elements within a system, analysts will be able to increase the number of parallel variables or constraints used to solve the problem [202].

*"I think there is a world market for possibly only five computers
[248]."*

Thomas Watson

# 8

# Agent Oriented Programming

An agent may be described as anything physical, synthetic or coded that is perceived
of being capable of interacting upon an environment. A human agent would be seen
to have *sensors* (eyes, ears, and other organs) to create *percepts* of the environment
and *effectors* (hands, legs, mouth, and other body parts) to act upon the environment.
A robotic agent may substitute cameras or other sensors to perceive the current situation, while various mechanised attachments could be used to effect some action
within that environment. When defining an agent, researchers describe the properties it should exhibit. The first property is *autonomy*, which means operating without
the direct intervention of humans. Second is *social ability* which describes the ability
to interact with other agents, agent applications and/or even humans. Third is *reactivity*, which includes a means of perceiving the environment and responding to any
changes that occur within it at a given point in time. Finally, *pro-activeness* means
exhibiting goal-directed behavior [407]. There are many architectures domains of influences and technologies that embody agent systems. When implemented as a system, agents are capable of achieving highly sophisticated goals autonomously and if
written correctly, will continue to find a solution until the goal is complete. This chapter investigates how agents architectures evolved, the level of control, construction
and mobility. Discussion continues to explore communications, how data is passed
or concepts merged and the technologies used.

## 8.1 Developing Agents

Before developing an agent one needs to understand what it is. Agents are generally
regarded as software systems and can be associated with an entity, framework,
architecture and even languages. Typically agents are piece of program code that

J.W. Tweedale & L.C. Jain: Embedded Automation in Human-Agent Environment, ALO 10, pp. 105–124.
springerlink.com

are able to autonomously complete tasks. They may be required to adapt, learn or collaborate with stimuli, sensors and actuators or data flows. This is generally done using some form of communication within or across a distributed systems or networks. The key arguments used to define agent include its context and in many cases, encompasses everything required to solve problems; past, present and future. The taxonomy or classification is important, however the concept of one entity being responsible to solve every goal presented is unrealistic.

## 8.2    Agent Intelligence

A simple study of our environment re-enforces the concept of *survival of the fittest*. Experts confine their efforts to solve problems within very narrow domains, which is an acceptable and efficient method of solving problems. There are associated costs and implications, but these constraints are tolerated. We need to force a paradigm shift to one that distinguishes the perception from the deliverables. The concept of one entity fixes all needs to be addressed and moderated to account for the technology at hand. Interoperable MAS using dynamic teams and functionality need to be considered. In order to achieve a flexible, reliable problem solving system, a multitude of functionality must be accessible to a coordinated and cooperative supervised team of agents or sub-systems. The concept of intelligence[1] used in agents may only be achieved by minimising the human element. In other words, provide software that achieves more than process or monitor without significant direction or control. Machines and production lines are controlled by operators that require specified skills to achieve a goal. When the required stimuli is missing or delayed, that machine or process become disrupted and may fail. The efficiency of attaining a goal, should not be confused with the intelligence of the machine or operator. Automation is the incarnation of a known sequence of series of processes that contribute to a predefined task. This concept should not be interpreted as intelligence, regardless of the level of technology or efficiency it provides.

As discussed, we know that humans are not efficient problem solvers because they suffer from cognitive stress when pressured to achieve. AI technologies can be used to personify many modern behavoirs or habits. Although agents are generally small and confined to limited functionality, they can be adapted to efficiently process specific tasks. Having a repository of such capabilities enables a collection of skills to be used to solve one or more elements of a decomposed task or goal. Being able to do this dynamically will improve efficiency and reduce complexity. The use of remote transactions or queries also reduce network traffic and improve reliability, especially across an unstable connection.

---

[1] Intelligence infers the ability to think, postulate or even compose a thesis to a solution. This requires many skills, compliance to lots of rules and a significant level of expert knowledge in the problem Domain. An Intelligence Quotient measures a range of skills using a standardised rating. Psychologist can administer professionally engineered tests, such as the Wechsler Adult Intelligence Scale (WAIS) or the Raven's Progressive Matrices. What is measured and its relevance is the major issue clouding the terminology. Computers appear to achieve intelligent feats, but they are not intelligent, they are merely doing what they are programmed to do!

Spiders, Bots and Aggregators were network based developments that preceded agents, however all traverse the web to achieve their goals. Each form is commonly encountered by programmers developing World Wide Web (WWW) applications. In its simplest form, a bot is a net aware program that behaves like a query statement. A spider has a more specialized design that targets content from other sites creating a map of its content. An aggregator is a function that produces a compound object from a collection of others on behalf of one or more users. Agents however, are viewed as customized assistants that generally use AI techniques to achieve their goal [149].

### 8.2.1  Building Spiders

A *Spider* is a special bot that is mobile and builds a list of resources (including internal links such as *href* and *map* tags), as it traverses from site to site. Using these links, the structure of the site can be reproduced locally. To achieve this function, the spider needs to be recursive (see Listing 8.1), however non-recursive spiders are also required[2].

```
1  void myRecursion {
2     myRecursion();
3  }
```

**Listing 8.1** Example of a recursive method

A number of queues are created to process every link found in the target URL. Heaton describes how to construct a spider using an interface and *Spider class* [149]. The resulting application (GetSite) processes a URL, extracts all its sites internal links and outputs the results to a file called *spider.log*. The *Spider* can be constructed as an event driven (*Go_actionPerformed*) servlet/bean. Thus it can then be invoked by a background thread that is synchronized with a Java Data Base Connectivity (JDBC) table [152].

### 8.2.2  Building Bots

A *bot* is an internet aware, autonomous *Spider*. It can be focused to perform specific tasks, such as inform the user of changes within the environment. These applications operate in the background. First they take a snap shot of the environment, then they monitor any changes and report those of interest to the operator. They can suffer from changes in the environment, such as site changes or a redesign of specific web pages, especially URLs. Some sites contain specific types of web pages that will interact with special *bot* classes, called *Recognizers*. Some freight companies now use commercial bots to track the delivery of parcels.

The recognizer class must be flexible enough to support different page content, such as those seeking country of origin and specific link classes. A *Tracker* example

---

[2] Non-recursive Spiders are used to retain a list of pages that appear to contain the data being sought and another that shows an error that may have occurred.

is provided in Listing 10.4 of Heaton's book[3] [149]. This is used to test the *jsp* which will return a tracking number to the *bot* when complete. Other recognizers are used to retrieve categorized information sources.

### 8.2.3   Building Aggregators

An *Aggregator* is a collection of *bots* used to provide a consolidated response based on data collected from several similar sites. An example may be as simple as collecting weather data on several cities for display on a single listing (possibly an itinerary or weather map). *Aggregators* can be written to run online (*thin client application*) or offline (*platform based/thick client* application). Online *Aggregators* have been provided by Yahoo, Charles Schwab, Bank of America and a variety of financial service companies. *Intuit* provides an example of offline *Aggregators* called Quicken.

To create an *Aggregator* agent, the designer needs to assess the knowledge targeted and the response required. This generally involves manually retrieving such information, however upon examining all the links some short cuts or automation could be realised. The HTML page is parsed to a *getTemp()* method. The URL must be formed from one of the internal links parsed. If the city code exists in the retrieve data, a value for the temperature is returned. The class may be extended to get a number of value each day, or even the temperature values for a number of cities and aggregate the results in a customised form.

## 8.3   Agent Development

Agent development has attracted significant resources resulting in a ubiquitous spread of frameworks, languages and architectures. Some of the more common development frameworks include:

- Aglet mobile agents by IBM;
- CAST by Multidisciplinary Research Program of the University Research Initiative (MURI);
- Java Agent Development Environment (JADE) by Telecom Italia;
- Java Agent Compiler and Kernel (JACK) by Agent Oriented Software (AOS); and
- State, Operator And Result (SOAR) by the University of Michigan.

### 8.3.1   Alglets

Aglets were developed at the IBM Tokyo Research Laboratory (TRL) by Mitsuro Oshima and Danny Lange [214, 215]. There are embodied in a Java library and instantiate as mobile agents with security provided by the Java 2 Security Manager [212, 213]. A development kit has been posted on *Sourceforge*. It is currently used in the development of an on-line or virtual travel agency called TabiCan and has

---

[3] To run this example you need to have **Tomcat** or an equivalent *jsp* server running on the host machine.

evolved into a Massive Multi-agent System [414]. The IBM Aglets SDK (formerly Aglets Workbench) is an environment for programming mobile Internet agents in Java (see http://www.trl.ibm.co.jp/aglets/)

### 8.3.2 CAST

This architecture was originally proposed to simulate the interaction of human-in-the-loop (HIL) activities when working in teams. It embodies the concepts of team-work using Petri Nets in the form of *Shared Plans* or *Joint Intentions* [139] using Multi-Agent Logic Language for Encoding Network (MALLET) which is loosely based on LISP [78,208,244,369]. These rules are communicated using Dynamic Inter-Agent Rule Generation (DIARG) [102] using a back-chaining theorem-prover called Java Automatic Reasoning Engine (JARE).

A number of researchers have integrated RPD model with agents [418] to capture the decision making abilities of domain experts. Klein created a model that evolved in an agent environment under teamwork setting into the RPD-agent architecture [101]. Hanratty et al. shows the architecture of the RPD-agent [145]. These agents have been tested in a military command-and control simulations. Under normal time pressure, the human teams are required to make correct decisions about the potential threat. The team performance can suffer due to the lack of information sharing during time poor scenarios. This can result in incorrect decisions being made and can be helped by cognitive based agents.

The SMM consists of team processes, team structure, shared domain knowledge and information-needs graphs. The IMM stores mental attitudes held by agents. The information is constantly updated using sensor inputs and messages from agents.

The AM module is responsible for the decision-maker agents attentions on decision tasks. The Process Manager Module (PMM) ensures that all team members follow their intended plans [418].

### 8.3.3 JACK

The JACK intelligent agent software provided by AOS for developing platform for BDI agent [7]. Here in this section we will address the enhanced model such as *TEAMS*. It focuses only on agent oriented teaming and their environment. The team model extends the agent concepts by associating the task and roles and coordinated behaviour. The team members instantiated in the teams model coordinate with each other to achieve given goals by keeping individual responsibility required to determine how to satisfy these goals. The JACK *Team* extension introduces the new concepts of team, role, team-data and team-plan. The JACK *Teams* Model includes all the programming elements of the JACK but with an extended semantics for some elements.

JACK Intelligent agents is a development platform for creating practical reasoning agents in the Java language using BDI reasoning constructs. It allows the designer to use all features of Java as well as a number of specific agent extensions. Any source code written using JACK extensions is automatically compiled into regular Java code before being executed. Each agent has beliefs about the world, events to respond reactively, goals that it desires to achieve, and plans that define what to do. JACK agents are based on the BDI reasoning model; it is Goal-directed, is Context sensitive and action oriented [161].

JACK *Teams* is an extension to the JACK platform that provides a team-oriented modelling framework. The *Teams* extension introduces the Team reasoning entity that encapsulates teaming behaviour and roles for defining what an agent is required to do. Belief propagation allows beliefs to be shared between members of a team. This means it becomes possible for sub-teams inheriting beliefs with important information from higher-level teams and conversely, enclosing teams to synthesize beliefs from lower-level sub-teams. JACK teams was developed to support structured teams, therefore the role obligation structure of a team must be defined at compile-time. Consequently, sub-teams can only communicate and share information if it has been previously defined in their team structure. KES conducted some collaborative work with AOS and implemented a simulation platform that could be used by students. This was used to simulate the collaborative and cooperative capabilities of UAVs [391].

### 8.3.4   JADE

This Framework is a free distribution provided by Telecom Italia [206]. It simplifies the implementation of multi-agent systems through a middle-ware that complies with the FIPA specifications and through a set of graphical tools that supports the debugging and deployment phases. As suggested by its developers, the agents can be distributed across machines as a language independent (using the JVM) application using a remote GUI. The configuration is dynamic during run-time where agents also have mobility when required. JADE is implemented in Java language (with support beginning at Java Run-time Environment (JRE) or JDK 1.4 environments) [385]. It can be integrated into your Integrated Development Environment (IDE) with a plug-in for Eclipse. The synergy between the JADE platform and the Lightweight Extensible Agent Platform (LEAP) libraries now allows the developer to obtain a FIPA compliant agent compatibility with mobility. This allows applications to execute across a wide range of devices varying from servers to Java enabled cell phones. KES has recently experimented with JADE to model robotic interaction in a hostile environment.

### 8.3.5   SOAR

Soar is a general cognitive architecture for developing systems that exhibit intelligent behavior. It is well documented [162, 210, 285]; however, Tambe extended this framework to incorporate Joint-Intentions via SharedPlans he called STEAM. It is a general model of teamwork, intended to enable agents to participate in coherent teamwork. Team operators are based on the joint intentions framework [61, 343]. It has been used to examine coherence within Teams as pilot agents in helicopter simulations for both *offensive* and *transportation* roles. Students have also used it to enter the *RoboCup* tournament held at Nagoya, Japan in 1997. Further documentation is provided by Tambe [360] and the SOAR website.

### 8.3.6  Building Agents

*Spiders*, Bots and *Aggregators* provide automated assistance to users looking for specific information. Future enhancements of these applications will increase their productivity and reduce the manual supervision/contribution required to achieve the tasks being conducted. It is feasible to extend these applications using the new packages and capabilities of Java or other languages under development. SOAP for instance, is a package that will enable sites to exchange data more easily as issues relating to semantics and ontology become less onerous [97]. HTML [302] is giving way to XHTML [291], empowering page authors to embody data using more powerful avenues of presentation[4] and new standards such as WSDL and OWL-S [43]. A number of autonomous agents now employ *bots* to provide weather, travel and financial services and this convenience is stimulating industry to raise the benchmark further. Hence the research effort towards MAS applications with mobile, distributed capabilities. The material covered in 'Constructing Intelligent agents Using JAVA' [25] will further extend the knowledge and practical skills required to understand the background of this development. Researchers will progressively perfect techniques to conduct reasoning, adaption, cognitive deduction, learning and goal oriented tasks processing in real-time. Given the laboured history of evolution, commercial interest is intermittent.

There has been limited commercial interest, reducing the number of possible success stories, however, research and development continues to evolve because the rewards are significant. The focus has been shifted to collaborative research, with many projects evolving as open source developments. The use of persistence, mobility and load-balancing are beginning to benefit from such activities. Mobile agents can now serve to reduce network bandwidth, application reliability and system performance measures. They facilitate more flexible designs, which enables reuse, adaptation and entities that support definable contexts. Due to the complexity surrounding this researcher, the number of collaborative projects have increased dramatically. Given teams of developers have specialist skills, it is important for everyone to understand the basic concepts.

Getting agents to *reason* or *make decisions*, shouldn't be confused with the ability to think or even its level of intelligence[5]. They can appear smart, mimic experts and portray knowledge using KBSs, but AI is being misrepresented by cinema and this image is still a distant vision (subject to scientific or commercial research and development). At present, supporters should project the terminology of CI [5].

Finally a 'Smart agent' is:

> "An agent systems that is truly smart, which would learn and react and/or interact with its external environment. In our view, agents are (or should be) disembodied segments of knowledge, embedded to represent intelligence. Though, we will not attempt to define what intelligence is, we maintain that a key attribute of any intelligent being is

---

[4] Such as direct XML calls via Java API XML Messaging (JAXM) [322].

[5] Procedural solutions need to provide the *what* and *how*, which when expressed as logic, can and does get very complex.

its ability to learn and that learning may also take the form of increased performance over time [280]."

An architecture provides the outline or basic structure of an agent instantiated at run-time. Using the human body as an analogy, this represents the skeleton, organs, muscles and epidermis. The problem with many existing programs, is that they don't possess the ability to transform the context or functionality dynamically or at run-time. Again, this could be seen as applying cosmetics or plastic surgery. The problem manifests using inheritance. Static links are embedded during compilations, restricting the adoption of new behaviours during run-time (dynamically). Although it is possible to embed the logic and actions required to address a specific number of predefined rules, it is difficult to modify any code while it is executing (at run-time), regardless of the conditions meet. Therefore we can only modify behaviour by redirection or branching logic without interrupting the current session. This compromises any existing state and or behavioural gains without complex entity management. The behaviour of an agent is effected in many ways. Autonomy makes it difficult to assess or react to the events an entity is subjected to in an external environment. Before understanding what needs to be done, it is important to understand the definition or expectations of autonomous activities.

Autonomy occurs across a variety of levels. The level of control is generally based on the functionality and response required. As agents encounter constraints or are deprived or resources, external influence may be required in order to maintain milestones to achieve goals. Therefore level of dependency needs to be balanced to ensure the whole system benefits. To achieve independence an agent must govern its own actions, subject to rules provided by its organic parent. Typical examples of commercial applications represent:

- Brockers,
- Auctioneers,
- Weather,
- Reservation, and
- Booking agents.

## 8.4 Mobility

The mobility characteristic was originally introduced into Agency technology to address issues related to distributed networking. This concept enables the agent to do the processing remotely (on the server machine) with the agent returning to the client when need for external stimuli/service has expired (similar to stored procedure transaction conducted on database engines on a server). Mobile agents evolved primarily to reduce:

- network bandwidth issues;
- design risks and glitches caused after amendment; and
- delays caused by intermittent, overloaded and unreliable networks.

Mobility enables the transfer of *state* and *logical resources*, both required by the agent to achieve goals remotely. Movement of code between the *client* and the *server* in lieu of streaming data or associated stimuli across a network for processing can significantly reduce network traffic, especially if all that is returned is the results (This could be a condition, value or an object). Unlike data transactions, an agent has the ability to *initiate/start, hibernate/pause and terminate/stop*. This process will be requested from the client (*remote host*) and reside on the server (*local host*) or peer until complete. A good example includes the concept of providing the measure of an aggregate sample of remote events. For instance, the mean measurement of the number of remote events or even the given frequency of that sample over a preset period (sampled from an addressable source). This concept facilitates remote, ad-hoc or customised queries without the need to create the required service or host globally.

## 8.5 Communication Methodology

All systems need to communicate in order to maintain synergy and sustain outputs. Programs can use a number of methods to communicate. These include; RPCs[6], callbacks[7] and event messaging[8] to exchange both data and intent. Java utilizes RPC for RMI, where you use callbacks within the AWT and Swing to control GUI objects. Event messaging relies on *reflection* and *serialisation* within **Bean** objects, to maintain asynchronous execution. The serialisation provides a unique identification for each entity and make packaging, segment transmission and reconstruction of the data stream between hosts. These techniques make messaging more reliable, the associated agent more robust and its migration almost seamless. The biggest issue with mobility is security. Hence there is a need to isolate external sources. To achieve this, the client needs to split the process to facilitate any redirection required. This could be achieved using a registry that lists the functionality available. Each message event must also be categorised and sent using a specified queue. The message must have a unique ID which is passed by providing a handle to its queue.

## 8.6 Passing Concept

If you understand the concept of how a thread or applet executes, an agent is also initialised (once and only once), while it may start/stop[9] as required to until the task concludes (once and only once). Agents need to be autonomous to succeed in a remote environment. It is acknowledged that nobody likes mundane, repetitive tasks, let alone having to mindlessly supervise/control something that does the same job.

---

[6] Where the execution is inherently sequential and prone to blockage/stalled execution.

[7] Where this call is asynchronous making the program more complex.

[8] Where the execution is also asynchronous and more suited to disruptive environments.

[9] Move, pause or terminate.

## 8.6.1 Parsing

HTML is a marked-up language that is composed of text, comments and tags [302] which itself is a subset of its predecessor SGML [63]. A more advanced XML Schema was created to express a shared vocabularies that enables machines to carry out rules made by people. This includes the structure, content and semantics required to perform that exchange [346]. A new protocol called XHTML is being developed that is being designed to integrate the best of both these schemes [291]. Some HTML pages contain Java Script (JS)(s) which can be used to control the content of a web page. Microsoft created DHTML [134] by embedding JS into HTML, while Sun created JavaServer Pages (JSP)s which could dynamically statically hold or dynamically update data between the client and server [237].

```
private static char[] toBase64 =
{  'A', 'B', 'C', 'D', 'E', 'F', 'G', 'H',
   'I', 'J', 'K', 'L', 'M', 'N', 'O', 'P',
   'Q', 'R', 'S', 'T', 'U', 'V', 'W', 'X',
   'Y', 'Z', 'a', 'b', 'c', 'd', 'e', 'f',
   'g', 'h', 'i', 'j', 'k', 'l', 'm', 'n',
   'o', 'p', 'q', 'r', 's', 't', 'u', 'v',
   'w', 'x', 'y', 'z', '0', '1', '2', '3',
   '4', '5', '6', '7', '8', '9', '+', '/' };
```

**Fig. 8.1** Allowable characters for Base64 [29]

Regardless of the protocol, embedded code is normally processed (*parsed*) using an interface written by the programmer. The Swing class provides a parser that is DTD driven. It is capable of parsing much more than plain HTML tags. A reader class can employ the *ParserDelegator* and use *ParserCallback* to manage the state during the process. This is normally achieved by over-riding predefined methods, such as:

- *handleText.*
- *handleComment,*
- *handleStartTag,*
- *handleEndTag,*
- *handleSimpleTag,*
- *handleError,* and
- *handleEndOfLineString.*

The HTML.Tag class currently lists 75 tags that act as constants and 80 more that represent attributes. Listing 8.2 provides an example of a callback class that could be used to parse the text within a tagged text document.

```
1  HTMLEditorKit.ParserCallback callback =
2    new HTMLEditorKit.ParserCallback () {
3      public void handleText(char[] data, int pos) {
4      System.out.println(data);
5      }
6  };
7  Reader reader = new FileReader("myFile.html");
8  new ParserDelegator().parse(reader, callback, false);
```

**Listing 8.2** Allowable characters for Base64 [29]


## 8.7 NASA Approach

NASA needed a cheaper method of designing, launching and maintaining scientific endeavours to survive. Such projects were vital to continuing a sustainable presence in space. Fully manned missions were no longer feasible; therefore the creation of a heterogeneous fleet of robotic explorers became the answer. These robots needed to survive very harsh environments with little guidance or remote control. Hence the use of remote agents that are embedded with CI. Model-based, autonomous agents, that contained the ability to deduce, search and achieve goal-directed goals were created and embedded into the space-craft mission systems [262]. The functionality required to operate during long periods of black-out, deal with tight schedules, resource constraints and concurrent activities in a tightly coupled system. The science fiction of technology has failed to materialize and as time passes we are still without a moon-base or HAL9000. Even the *International Space Station (ISS)* is a modest creation of its original vision. The first candidate of remote agents was *Deep Space One (DS1)* and the challenge chosen was to implement a tractable reasoning and knowledge representation system. Due to the long term communication disruptions and message turn-around times, the virtual human presence normally present may be represented in the absence of such *televisory* command and control.


### 8.7.1 The Paradigm Shift

Previously humans have maintained a *televisory* link in the decision loop, although due to intermittent communications, temporal response and the need for immediate control decisions, celestial travel has created the need for a *virtual human presence*. This has led to the exploitation of remote agents using AI conventions to implement embedded systems with tractable reasoning and representation [262]. The shift has occurred for many reasons, however primarily due to design and maintenance cost. The Galileo mission cost over US $1B and requires a ground control/maintenance crew of between 100 - 300 people, 24 hours per day for the whole of its mission. Budget pressures offered no alternatives, go lightweight or abandon the flight program altogether. The design of the Mars Pathfinder (MPF) mission needed to show an order of magnitude in savings before being approved. This created a significant challenge for scientists because all previous on-board systems had no embedded intelligence

(A significant cost in maintaining such projects). More ambitious projects require teams of distributed intelligent agents to operate remotely, such as the Cryobot Probe and Hydrobot Submercible, DS3 and the Martian plane. To cut costs, design times also had to decrease, therefore the plan for DS1 was reduced to 30 months and was required to demonstrate:

- long term autonomous operation;
- guaranteed success based on tight deadlines and resource constraints;
- exhibit high reliability; and
- conduct concurrent activities within tightly coupled subsystems.

To achieve the required autonomy, a New Millennium Autonomy Architecture Prototype (NewMAAP) agent was developed by a group of AI researchers in six months at NASA Ames and Jet Propulsion Laboratories (JPL). The new agent contained "integrated constraint-based planning and scheduling, robust multi-threaded execution, within a model-based identification and reconfiguration module" [262]. Three distinctive features prevailed:

- A model-based module using programmable composition;
- The ability to perform embedded *deduction and search*; and
- Provide a *higher level closed-loop command* capacity.

This style of programming is based on observation using a breadth first approach to accommodate the functionality required in a given set of hardware. A best-first search algorithm was embedded into the kernel [400] to cater for a wide range of narrowly focused diagnostic plans. This was implemented in a Reduced Instruction Set Computer (RISC) approach to multithreaded goal directed execution. This avoids the creation of a brittle *mission profile* across parallel mission segments[10].

## 8.8   Other Effects

Cognitive Science is a field of research attracting significant effort. It was preceded by the process management evolution with many prominent achievements, such as, Taylor's introduction to Scientific Management and the Hawthorn Experiments conducted by the National Research Council (NCR). Formalising organisational systems and behavioural science provides the tools required to decompose human oriented task. Any real-world system takes inputs as sensors will only react appropriately when it is able to modify the outputs. Simulation models rely on the same approach. Agents can be used to monitor sensors and stimulate the decision making required to modify one or more outputs.

---

[10] Apollo 13 experienced a quintuple fault which required an army of ground crew the challenge of remotely assessing the status/health of the spacecraft prior to rapidly redesigning a new mission plan with revised procedures. The ultimate decision compromised the original mission goal of landing on the moon which was quickly revised to a successful return to earth. The ground crew were required to search for a new unintended reconfiguration of the space crafts subsystems with the required procedures required to effect those changes manually.

Agent technologies, and in particular agent teaming, are increasingly being used to aid in the design of "intelligent" systems [380, 411]. In the majority of the agent-based software currently being produced, the structure of agent teams have been reliant on that defined by the programmer or software engineer. The development of a model that extends the communications architecture of an agent framework that is adaptable when contacting a series of MAS or teams. The ideal properties of agents, includes: deliberative agents, reactive agents, interface agents (HCI) and mobile agents[11] [95]. Different systems may be instantiated with a variety of hierarchies, with each level performing predetermined tasks in a subordinate or supervisory role. An Agent Architecture is considered to include at least one agent that is independent or a reactive/proactive entity and conceptually contains functions required for perception, reasoning and decision. The architecture specifies how the various parts of an agent can be assembled to accomplish a specific set of actions to achieve the systems goals. Wooldridge believes that it is essential for an agent to have "the ability to interact with other agents via some communication language [410]".

This research requires the formation of teams of agents in order to dynamically configure the team with the ability to solve the decomposed task of the goal presented. Traditionally all tasks must be completed successfully or the team fails the goal [86, 411]. A dynamic architecture would substitute agents within the team with alternative capabilities in order to succeed. It may even compromise and offer a partial solution and offer it to another system to complete. A good communications framework is required to pass messages between separate agent and other systems. Discussion about confined frameworks have recently been extended to enable individual students associated with our KES group to fast track the development of their research concepts. A *Plug 'n' Play* concept based on a multi-agent blackboard architecture forms the basis of this research. We believe a core architecture is required for MAS developers to achieve flexibility. The research focuses on how agents can be teamed to provide the ability to adapt and dynamically organise the required functionality to automate in a team environment. The model is conceptual and is proposed initially as a blackboard model, where each element represents a block of functionality required to automate a process in order to complete a specific task. Discussion is limited to the formative work within the foundation layers of that framework.

## 8.9 Remote Procedure Calls

A remote procedure call is a function or subroutine that enables inter-process communication. Modern examples of client/server communication use the traditional *stub* and *skeleton* framework. There are several popular implementations with IEEE, ISO and DCE specifications. Synchronous threads enable protocols to traverse from *Application* through to the *Transport* layers across distributed networks. Stubs have been written in many languages and is now embedded in a variety of frameworks. When using distributed systems, middleware was used to manage transaction processing. Using Corba Object Request Brokers (ORBs) provided an interface that enabled a standardised IDL to pass messages using the Internet inter-ORB Protocol (IIOP) in

---

[11] Inter and intra-net.

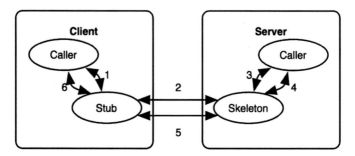

**Fig. 8.2** Remote Procedure Call [358]

the manner depicted in Figure 8.2. Similarly, RMI and SOAP are used within Java to provide this capability.

## 8.10   Platform Migration

The concept of an ORB has been some what outdated with web based applications and services. Mobility, distribution and concurrency are no longer confined to middleware such as Corba. The JVM and Java Run-time Environment (JRE) can now be used to provide the same type of functionality. Traditional resource management creates dependancies, latency and imposes vendor designed constraints. Desktops are now being treated as virtual entities using a namespace model based on URLs. In Java this can be demonstrated using RMI, Representational State Transfer (REST) and SOAP. This is a developing field of research and new examples are being shared, especially through open source projects. Performance and efficiency of systems across network observes the underlying concept of *clusters*. Given the concept of *platform independence* the hardware should behave as a transparent element to any software. The burden of resource management has traditionally fallen on the developer, with most users being unaware of how, when or on which platform the actual software has/is being executed[12]. A number of techniques have been developed. These include:

- skeleton/stub connections (RPC),
- post and wait,
- callback, and
- event-based messaging[13].

Event queues are becoming commonlace in modern applications. Components and services are being assembled to provided functionality using Model View Control (MVC) patterns. Software designs have adapted, by separating *Business Logic*, *Controller Logic* and *System State*. At the same time, communications models have

---

[12] They do have influence on the where and why.

[13] This form of call is analogous to hardware interrupts, were a solution is posted to service specified events.

evolved to facilitate this form of reuse and interoperability [80, 316]. Traditionally this process relied on developers, required specialist skills and resources and tended to be proprietary in nature. Such processes were developed as middleware and implemented as services within the sponsor organisation.

## 8.11 Service Oriented Architecture

The structure of a service must provide the cross cutting functionality, while controlling the session. It provides the domain logic in a stateless environment, using a local interface which is deployed with mandatory settings. Most enterprise applications are based on SOA. Java implements *generics* to fashion SOA interfaces used in enterprise architectures. The structure of a service must provide the cross cutting functionality, while controlling the session. This is a huge topic and we only discuss the metaphor and its influence in the public domain.

## 8.12 Models

Designers commonly mix metaphors when describing their designs. An application, function or capability can be loosely described using the term model, framework and architecture[14]. A number of paradigms are used to assist designers document their goals or what they wish to achieve. It is important to divide the project into its constituent components. The model should contain the information or state, where the view represents what is being managed (commonly via the user interface) and the application logic or controls are used to map the user response or stimulating actions. It is possible to use UML to document Plain Old Java Object (POJO) classes, include the binding relationships, the interfaces used and the event processing considered at compile-time. A number of models may be used within a single framework[15]. In a more complex system, a number of frameworks may be incorporated into a larger architecture in an attempt to solve real-world problems. Given that declarative binding is static it is difficult to make dynamic variation at run-time without the developer embedding some form of fairly complex or detailed decision algorithms to resolve any unforeseen conflicts to prevent system failure[16]. This task has been simplified in Java, especially when programming distributed applications, with the inclusion of additional annotations, including *@Namespace, @Transform* and *@Refreshable*. By changing the logic within the dispatcher it is possible to transform a local application (an agent) into one that is portable (*mobile*) without changing the business logic that created it.

---

[14] Models can also be confused with the concept of modelling and simulation.

[15] It is interesting to note that Patterns have emerged in the form of fully documented framework that implement common programming concepts.

[16] This requires more resources at design-time, increases complexity, creates complex logical anomalies and impedes re-use/re-design.

## 8.13   Patterns

The GoF introduced the concept of design patterns to the general community. Patterns capture the experience of others and promote re-use. They help programmers to communicate, implement and maintain aspects of software design. Java provides examples of this approach when implementing the AWT (*Strategy Pattern*), User Interfaces (UIs) (*Composite Pattern*) and I/O (*Decorator Pattern*). A container can modify its layout without the need to make changes to the code elsewhere, thus encapsulating the core concept, making the implementation less complex, easier to modify, extend and maintain.

In many applications, communications can become disruptive and cause delays. This is generally due to the sequential implementation of protocols and serialisation of traffic streaming (delays effectively block the execution of some events). Programmers need to provide modular or segregated implementation that is preferably concurrency using independent business logic. Components can be added in an aggregate fashion to compose more complex structures, while stream readers can be decorated using the *java.io* package to configure or harness messaging and event queues. Using an analogy based on the human body provides a good example. The body has many sub-systems in the form of organs. When the external environment influences an organ, it may be necessary to react. If the effect is severe, the organ doesnt stall or wait for resources to respond. They are symbiotic and embedded within other body sub-systems such as the nervous system[17]. In fact the human body is composed of eleven sub-systems. These include the: respiratory, digestive, muscular, immune, circulatory, digestive, skeletal, endocrine, urinary, integumentary and reproductive systems [216]. Evolution enables complex systems to interoperate, however communication is central to how all organisations operate. Occasionally the intentions of many organisations are interrupted by environmental factors. Sensors are used to harness these effects using sight, sound, taste, touch and smell.

When developing an Object, it is possible to use a core component, such as a table and extend this using inheritance or polymorphism or simply decorate the base object using a series of implementations that provide additional functionality or look and feel. Wrappers reduce duplication and enable a recursive approach to decorating objects. Examine the implementation of the core *Buffered File Reader*. A single call can create a *Buffered File Reader* using **BufferedReader** *bfrdr = new BufferedReader(new FileReader(new File(filename.ext)))*. If you needed to access a specific line, this could be further wrapped using the LineNumberReader class. Again, developers can drill down further by tokenising the line read. A similar approach could be used to wrap dynamic behaviour or decorators to a Swing Table [124]. The container can be created to display the abstract Table and represent its content (data) based on a user's preferences. Events could be captured and processed to sort (using the defined default or an algorithm dynamically selected), filter or modify the contents at run-time without the need to re-write any code.

---

[17] An electrical system runs parallel to the circulatory system.

## 8.14   Proxies

The proxy design pattern illustrates how an item can be represented (indirectly) in one location by mimicking the reference of another object without being copied. In Java a proxy can be used to deliver *Web Services* on another machine by passing a unique URL via the Java API for XML based RPC (JAX-RPC). This approach of accessing the *behaviour* of a distributed service is generally termed SOAP and uses WSDL. It is sometimes confused with REST which transfer an objects *state*. They are fundamentally different technologies, SOAP can be abused to do RESTfull things [80]. Three Java classes support proxies are based in the *java.lang.reflect* package. These classes include; *Proxy, Method* and *InvocationHandler* which are more commonly encountered when creating *Bean* objects (moving from loosely coupled *whitebox* implementations using inheritance, to tightly coupled *black-box* re-use through composition. The major difference between a decorator and proxy pattern is the *intent*. As discussed, decorators are applied recursively at run-time, where proxies merely used as stand-in to duplicate the original object.

## 8.15   Generics

Generics are being introduced into Java to provide a stricter level degree of *type safety* and assists in creating parameterized types (reducing the requirement to add significant amounts of casting and error checking). Generics will also have an effect on the way in which *varargs, annotations, enumerations, collections* and even how *concurrency* can be implemented. Generics don't apply to primitive type, but that doesn't mean those types cannot be parametrized. The key to understanding Generics is to recognize that they form a hierarchy based on *type* NOT the *parameter(s)* passed by that type [250]. The advantage of Generics is the ability of using wild cards without parameterization, so when creating a new instance of a specified type its a simple matter of substituting the placeholder (type) with a specific type like; String, List, Map, Container, similar to templates in C++. If this can be achieved at run-time the concept of a type safe dynamic function with a variable number of arguments becomes a reality. At present we can use Generics to reduce coding complexity while maintaining safety. However a filtering function must be used to indicate the type being used at any point in time with specific sub-classes written to support those types. Although Generics are not the central focus of this text, they are a technology that can provide a way forward for the concepts raised.

## 8.16   Technology

Developers maximise new technology as it becomes available. Over the past decade 64-bit processors have become main stream, with new CPU designs and multi-core processors are now becoming common place. These new chips have facilitated the supply of desktop machines that are capable of true parallel processing and direct memory address that increases performance using relatively less power, without network bottlenecks commonly encountered with distributed systems. Examples of advancement include the Intel *Westmere, Nehalem, Gulftown* and *Tukwila* CPUs [40], while Advanced Micro Devices (AMD) introduces the *Deneb, Orchi, Camplain* and *Liano* microprocessors [259]. Sun and IBM have been serious about parallel processing for many years. Sun and its open source development community that supports Java have turned their attention to multi-core issues by incorporating optimisation into the virtual machine and the Java Developers Kit (JDK). Concurrency has been supported since 2004 enabling programmers to execute threads in parallel. Sun is reported to release a new type of garbage collection and enhanced concurrency in JDK 7 due for release in 2010 [253]. The recent Mustang release (Java SE 6) also provides enhanced scripting, application management Java Management Extension (JMX), web services Java XMLWeb Services (JAX-WS) 2.0 and Java Authentication and Authorisation Service (JAAS)) and desktop improvements (JDesktop Integration Components (JDIC)).

## 8.17  Threads

Much has been written about threads over the past decades [223], however as multiprocessor silicon becomes available to the desktop, the topic has grown in popularity. Carver revives the theories of Tai [315] and has coded examples to support his arguments. The concept of multi-threaded applications requires a library of support functions to assist in the control and successful execution of user provided applications/threads. POSIX has become a unified standard (IEEE 1003.1) that provides this library which could be used to reduce the differences between multi-threading capable operating systems. The concept of Multi-Threaded (*OS/2, NT, Solaris and Unix*), Multi-Processor (*SPARC*), Symmetric Multi-Processor (*CRAY*) and distributed operating systems (*CORBA and Spring*) will also be discussed.

Threads provide programmers with a powerful tool to enhance the interactivity of a graphical environment. Java provides a stack-based[18] Thread class that includes methods to start, stop, run and check a threads status. These threads may be synchronised using a paradigm based on a set of sophisticated synchronisation primitives, postulated by Hoare [155] twenty year ago, implemented on the Xerox PARCs Cedar/Mesa system. Java provides threads that are pre-emptive (based on time-slices) and includes a yield() command to enable threads with higher priorities to yield control to lower priority threads. Classes may be declared using the synchronized keyword, however they do not run concurrently. They are controlled using re-entrant monitor variables to ensure that all object variables remain in a consistent state when being switching [135].

Lewis and Berg [420] provide a detailed discussion on Multi-threading in relation to *Solaris, OS/2, NT* and *POSIX*. They cover synchronization and locks, in-

---

[18] In Java a stack frame consists of three (possibly empty) sets of data; the local variables (for method calls), its execution environment, and its operand stack. The size of the first two stacks are fixed at the start of a method call, where the operand stack varies in size as bytecode is executed within the method [221].

cluding: *"mutexes, conditional variables, read/write locks and semaphore"*. This discussion is augment with a description of *"spin-locks, dead-locks, races, priority inversion and reentrancy"*. The POSIX IEEE 1003.1 Standard is also introduced, presenting a question on the effectiveness of monolithic operating systems designed using traditional top-down programming techniques. Especially when a collection of multi-threaded components that have dynamic capability could prove more useful. Given the proliferation of multi-core microprocessors from Intel/AMD and GPUs from Nvidia, threaded agents are becoming common place. Techniques previously reliant on High Performance Computing (HPC) architectures, are also being migrated to desktop and mobile platforms. As such, they are now being integrated into non-enterprise agent based applications.

## 8.18   Agent Programming

It should be noted that agents are not new, neither is distributed processing or migration. Developers have created many novel approaches to implement decision making for use in agents. Many developers have also created frameworks and larger architectures by combining or incorporating existing techniques. These techniques could be used to support or enhance the resulting technology as it evolves. Parallel computing emerged in the mid 1950s and was confined to mainframe computers. These techniques have recently been adopted by main stream computing, using clusters or even personal computers with multi-core microprocessors. Parallel processing support is now appearing in common development languages, like C++ and Java. There is a significant amount of information surrounding AOP and we offer the reader the opportunity to review this topic at their leisure [286, 335]. Some of the issues driving design(s) include:

Resources:     The ability to solve the problem (or Goal) within the local system;

Environment:   The need to implement a distributed system and the need to separate each capability; and

Effort:        The use of Patterns that have been developed for each of these capabilities or that can be implemented as a wrapper that can dynamically modify the core design using a fascard or decorator concept.

## 8.19   Real-World Applications

Agents are used almost everywhere, but there is an increasing demand to create sophisticated (semi-intelligent) applications, in fields such as: Entertainment, Medical and Astronomy. Although there a numerous examples, some include:

Multi Platform Games:  OneTooFree's *Wurm*[19].
MMORPG:              Jagex's *Ruenscape 3D*[20].
Medicine:            The National Library of Medicines *Visible Human Project*[21].
Astronomy:           NASA's *JTrack 3D*[22].

## 8.20  Next

Agents can be programmed in many languages, they are found embedded in a variety
of architectures and essentially enhance the developers ability to solve problems. A
Java is being taught as a main stream language in a significant number of universities.
We will confine our descriptions of agency theory to that context. It is not perfect,
however is suitable to explain how an agent can be packaged and even attached to an
application in memory at run-time. Prior to describing this implementation, we need
to outline the problem domain. A typical problem domain used to describe AI theories
revolves around game theory. In this case, we have chosen to use the Sudoku Puzzle.
In the next chapter, we describe the background, rules and constraints involved.

---

[19] See http://www.wurmonline.com/

[20] See www.runescape.com/

[21] See this project at www.codehead.org/human

[22] Again see these resources shared to the community through
   http://techtran.msfc.nasa.gov/software/jtrack.html

*"Language is an instrument of human reason, and not merely a medium for the expression of thought ... [248]."*

G. Boole

# 9

# Creating an Agent Factory

In this chapter we describe the requirements needed to construct an Agent Factory. We discuss how many of the fundamental Java capabilities can now be used to build credible applications to solve enterprise style problems. A brief list of recent enhancements are provided with a number of future changes flagged.

## 9.1 What Are Agents

Researchers have witnessed several evolutionary phases of agency theory. These include; mechanical/analog, processes/threads and the transformation from primitive software daemons through to IA based on MAS. More recently, MAS have been constructed with mobility, agility and intelligence. Examples range from gyroscopes (once a large bulky machine originally used for ship navigation), processes and daemons that control simple capabilities (such as thermostats and printers), ranging through to highly integrated systems (like those found in aerospace or even in-car navigation systems). We understand that agents have a variety of definitions and use, however the capability and its wrapping mechanisms need to be treated separately.

## 9.2 Agent Capability

The capability performed by agents normally utilise existing theory, such as search, reduction, reasoning, pattern matching and even learning. Technologies embedded

J.W. Tweedale & L.C. Jain: Embedded Automation in Human-Agent Environment, ALO 10, pp. 125–144.
springerlink.com                                                    © Springer-Verlag Berlin Heidelberg 2011

within agents that delivers these concepts is not exhaustive and not limited to logic, neural nets, expert or knowledge-based systems. A basic language of functionality could be implemented to enable simple statements, used within context, that can solve problems using the same design.

## 9.3    Agent Wrapping Technology

This is generically the corporate glue that binds the capability and provides the implicit behavior inherent to an enterprise level environment. Like Java Tables that can be decorated to enhance its appearance and modify displayed behavior, agent capabilities can be wrapped with the complexities inherent to its environment. That wrapper can be customized, extended or modified to implement system dependent constraints. In this case we confine our discussion to techniques used within the Java language.

## 9.4    What Is Java

Java is a fairly complex language that has already endured numerous iterations. As defined by James Gosling in the inaugural Java *White Paper*, Java is a simple, object-oriented, distributed, robust, secure, architecture neutral, portable, high-performance, interpreted, multi-threaded, and dynamic language [135]. Many authors have since scrutinised Java and most agree that with each iteration, Java has surpassed its design goals. Most acknowledge the Object Oriented Programming focus, its cross platform capability and its ability to access legacy system data.

Essentially Java is marginally different to other languages. The programmer still produces source code that is compiled and run on the target platform. C/C++, Fortran, Pascal and ALGOL are all examples of this three step process. Most languages must be recompiled when run on different platforms or targeted to new operating systems. This creates distribution problems, version management issues and additional expense. To alleviate these problems Java creates an intermediate file called Byte Code (although it is possible to produce native binary code using the J2C compiler). This code is architecture neutral and requires an interpreter (termed a Virtual Machine) to execute the *Byte Code* [372].

Java removes some of the familiar C++ features such as pointers, pre-processing, multiple inheritance, goto, and automatic coercion [394]. It does however add features such as references, index safe arrays and automatic garbage collection. Gosling originally attempted to extend the C++ compiler, however realised that even with the extras, it was not enough. Today JIT compilation provides near real-time performance. Table 9.1 provides a brief comparison of several OOPLs followed by a brief history of its evolution.

**Table 9.1** Object-Oriented Language Comparison [44]

| Feature | C++ | Objective-C | Ada | Java |
|---|---|---|---|---|
| Encapsulation | Yes | Yes | Yes | Yes |
| Inheritance | Yes | Yes | No | Yes |
| Multiple Inh. | Yes | Yes | No | Interfaces |
| Polymorphism | Yes | Yes | Yes | Yes |
| Binding | Both | Both | Early | Late |
| Concurreny | Poor | Poor | Difficult | Yes |
| Garbage Colection | No | Yes | No | Yes |
| Genericity | Yes | No | Yes | Yes |
| Class Libraries | Yes | Yes | Limited | Yes |

## 9.5   Recent Java Enhancements

When Java was originally launched, the major issue revolved around its lack of ability to print. Subsequent releases provided enhanced graphical capabilities via the AWT and Swing libraries. Sun enlisted the user community to evolve Java and many C++ capabilities found there way into the Software Development Kit (SDK). The most recent enhancements were stimulated after Oracle bought Sun and stimulated the community program based on Java Specification Program (JSR) categories. Since RMI we have increasingly used Java for distributed computing and enterprise style agency development. Table 9.2 catalogs the branding of major influences as derived directly from Sun press releases. For a complete history, see *The Java History Timeline*[1].

| | |
|---|---|
| JDK 1.0: | This was the first public release of work original titled the *Green* project. It was reported as a *write-once-run-everywhere* capability. Its major feature included the *Java.awt.Event* (which used the Event Model). |
| JDK 1.1: | This release added inner classes, JavaBeans, JDBC, RMI and Reflection, while the AWT was also updated. The *Java.util.EventObject* (which uses the Delegation Event Model) received significant appreciation. |
| JDK 1.2: | Introduce *Collections*, JIT and an IDL implementation for Corba interoperability. |
| JDK 1.3: | Updated RMI to support Corba, sound, Java Naming and Directory Interface (JNDI) and Java Platform Debugger Architecture (JPDA). |
| JDK 1.4: | Introduced *assert*, New I/O (NIO), native XML, Java API for XML Processing (JAXP), cryptography and *Web Start*. There were several subversions released, with 1.4.2 where |

---

[1] At http://www.java.com/en/javahistory/timeline.jsp

**Table 9.2** Java release history

| Version | Codename | Released | Month |
|---------|----------|----------|-------|
| JDK1.0 | Oak | 1996 | Jan |
| JDK1.1 | JavaBeans[2] | 1997 | Feb |
| J2SE1.2 | Playground | 1998 | Dec |
| J2SE1.3 | Kestrel | 2000 | May |
| J2SE1.4 | Merlin | 2002 | Feb |
| J2SE5.0 | Tiger | 2004 | Sep |
| J2SE6.0 | Mustang | 2006 | Dec |
| J2SE7.0 | Dolphin[3] | 2007 | Jan |
| J2SE8.0 | OpenJDK[4] | 2012 | Mid |

the *Javax.swing.event* was updated (using the *EventListener* Model).

JDK 1.5:  This release represents a major revision and was given a new version number. Java2 (Version 5.0 - Tiger) introduced reflection into the library core, enabling the use of Meta data. New functionality that included: *genetics, concurrency, auto-boxing* and *enumerations* were added. Swing received a face lift (*skins*) as did collections (*for each*) and *Varags*. Java2 attracted a significant amount of editorial, the best by Robert Eckstein who described JSR 270 in detail [93]. The *Instrumentation Class* was also provided, but failed to attract the initial attention it deserved.

JDK 1.6:  Called Java2 (Version 6.0 - 'Mustang'). It was based on a *publish and subscribe* concept that enables observers to monitor a specified event. Public access to events is required, however it would be easier to implement if the scope was accessible via using a *service* or *queue*. There are no significant changes to the language or its structure, however support was added for pluggable annotations, scripting and JDBC 4. There are a raft of enhancements that include; core performance, XML support and desktop features. Changes to the collection provide interfaces called *Deque, BlockingDeque, NavigatableSet* and *NavigableMap*. Deque for

---

[2] No official name was given to release 1.1, however this version included JavaBeans[TM] improvements to the AWT, new JDBC features, Unicode support and a revised Application Foundation Classes (AFC).

[3] Java SE7 was started in Jan 2007 but suffered delays. It was re-scheduled for release some time in 2010 (A release date of Apr 2010 was reported, however this has been abandoned in favour of minor updates to Java SE6.).

[4] Has a planned release date of 2012. Essentially this will contain many of the components that fail delivery targets of SE7.

instance enables bi-directional traversal or an object. New classes include; *ArrayDeque, LinkedBlockingDeque, ConcurrentSkipListSet, ConcurrentSkipListMap, AbstractMap. SimpleEntry* and *AbstractMap.SimpleImmutableEntry* with updates to *LinkedList, Collections, TreeSet* and *TreeMap* [94]. The most significant change for Agents was the introduction of the ability to dynamically install agents at run-time using the instrumentation class. It also enables redirection and the ability to use the **IsAssignableForm**, especially helpful when conducting hierarchically aware transforms. Again the *Javax.enterprise.event.Event* was updated (using the Observer Pattern). Of more significance, it introduced the ability to dynamically install agents at run-time with the instrumentation class, it also enables redirection. Several reasons for modifying an Agent class when being loaded include the need to provide a discoverable *identification, log activities,* implement *semaphores* or to enforce customised *communication* protocols.

JDK 1.7: Again called Java2 (Version 7.0 - 'Dolphin'). At the time of writing had not been released, but is reported to include; new annotations, improved concurrency, enhanced class loader, stream control transmission protocol (Solaris), socket direct protocol (*infiniband*), better cryptography, XML stack and date support. The upgrade was originally reported to include features like the new Module System (Super Packages), Servlets (new NIO package), Spring 2.5 (specifically the Date and Time API) and better integration of the asynchronous Java Messaging System (JMS) with Spring (Cache API) [253].

JDK 1.8: Java SE8 has already been planned. JDK 1.8 will be based on JSR 337 and capture the components that failed to make release SE7. A planned release date has been set for 2012. It will focus on the effort currently being expended on projects Lambda and Jigsaw.

## 9.6 The Next Release - Java SE7

With each new release of Java, the ability to handle distributed autonomous programs that run in their own thread with the ability to communicate between each other becomes ever easier. The Agent class was first introduced in version SE5 and further refined in version SE6. The JMX provided the ability to *analyse* or *test* applications with agents or even applications running in separate JVMs. Java SE7 was due for release in early 2007, but has fallen prey to delay. A series of minor updates to Java SE6

have been released in the interim (23 by the end of 2010). Table 9.3 lists a number of
suggested improvements which are based on the following JSRs originally submitted
for industry concurrence (some may be issued as interim updates) [253]:

**Table 9.3** Notable JSR Activities

| JSR | Description |
|-----|-------------|
| 166 | Provides *folk* and *join* processes to support concurrency. It will also provide a set of work queues to support *work stealing* to manage idle workers. |
| 203 | Extends the work started in JDK 1.4 and will improve *File* and *Directory* permissions. |
| 255 | Provide JMX updates to enable the *federation* of remote services. |
| 270 | Use components in JTabbedPane. |
| 292 | Culminated discusion on *tail cal optimization* and *interface injection*, however may only see **invokedynamic** included. |
| 294 | This was raised to improved modularity and namespaces and became project *jigsaw* when the OSGi Alliance joined the project. |
| 295 | Is targeted to improve Swing and and Bean binding/validation. |
| 296 | Use SwingWorker with JFC & Swing. |
| 299 | Better Dependency management similar to that experienced with *Marven* and *OSGi*. |
| 310 | Provides better *Date* and *Time* dimensioning in an attempt to provide *socialized* labels which will be great for managing behaviour[5]. |
| 366 | Expand the multi-threaded and multi-core capabilities within the Fork/Join Framework. |

In Java the *Facsard pattern* is used to frame the scroll, spin, mouse, keyboard and move events within GUI components. These should be implemented by
default and enabled as required (without the complexity of public or abstracted
constructors). Obviously they should retain the ability to override or extend such
methods, but each object, bean or tool, should operate using a single call (to create a default object) with methods or alternative constructors that can be used to
customise its properties. Programmers should only need to override a default API
when an alternative is required. For instance a button generally always requires
an event handler. Most GUI programmers include mouse control (click events),
then modify their designs to include keyboard shortcuts (key events) and increasingly some even add finger control (touch events). Icons, Actions and Default
values should simply be passed during construction and work without additional
effort (customisation would be achieved with additional method). If this were the
default condition, many aspiring programmers would achieve a more meaningful contribution with less complexity and frustration. The argument over reuse
and flexibility would still be retained by using interfaces, wrappers or redirection.

---

[5] This will provide methods to manage Episodic, Scientific, Empirical and System time measurement.

Why not rely on existing implementations. Traditionally visibility is achieved using inheritance which bloats code. Either that or the prolific use of predefined redirection using the *instanceof* call in a complex messaging system based on sockets or queues. Presently developers can use **invokeLater** to queue events or **invokeAndWait** but this compromises asynchronous implementations. Using the SwingWorker **doInBackround** method, designers can create a *Property Change Listener* thread that subscribes to interoperate with a GUI thread (usually a component).

Annotations[6] were introduced to Java at version 5.0 and improved/extended at version 6 to facilitate the use of *meta-data*[7]. Annotations can be used to provide: information for the compiler, conduct compiler-time and deployment-time processing or to enable run-time processing. An Annotation can be declared as a class, field, method or other program element. The annotation processors generates the code you need in your day-to-day programming at build-time[8]. Originally, Java programs were 'decorated' with annotations indicating which methods were remotely accessible, especially when writing JAX-RPC web service and when adding Reflection. Enterprise Java Beanss (EJBs) also required a *deployment descriptor* to reduce complexity and enforce its strict *syntax* requirements. Annotations have proven popular in Java and many tutorials now exist, therefore it will not be the subject of discussion in this book.

The trend in simulation applications is to 'Separate' the scenario data from the business logic. Traditionally the business logic is programmed to represent both the environment structure and its behaviour. This concept needs to be further extended to simplify an implementation that supports a dynamic context. The new model should retain the business logic that supports common the functionality (model kernel), but separate the context and behaviour. A XML Object Model (XOM) file provides the mechanisms required to encode the behaviour of the changing context[9] within the base environment, the environment itself and each scenario instantiated.

Existing simulation frameworks are more complex than necessary because they need to address all possible events[11] for any given entity that interacts with the environment. *Behaviour*, *desire* and *intent* are generally infused with the business logic. Alternatively the implementation logic inherits a communication model, such as KQML, ACL or FIPA. SOAP is emerging as the model of choice for distributed communications in Java.

---

[6] JSR 175 - A Metadata Facility for the Java[TM] Programming Language.

[7] @Annotations were originally implemented to enable developers to introduce *tags* or their own pre-processes, similar to that in other languages.

[8] This is sometimes called boilerplate code and could include support for strict syntax encoding, the generation of XML files or even documentation (like *javadoc*).

[9] Java makes use of the *java.util.Observable* pattern to implement Java Contexts and Dependency Injection for the Java EE platform (CDI)[10] to *weld* components using the JMS.

[11] Events contain objects or messages that are used when a software component(s) wants to notify a state of change to another component(s) within any package, thread or application.

## 9.7   Java MAS Framework

Lewis discussed the use of Object Oriented techniques to improve reuse [225], while Johnson introduced the concept of using a skeletal approach to constructing frameworks [180]. There is still a significant amount of research within the AI community on MAS, especially across distributed systems. Java has long been regarded as a learning language, that is slow, it leaks, its abstracted at a high-level, doesn't support gaming, was stifled by Sun and not appropriate as a development language by the commercial world [78]. Since then Open source development communities have made significant contributions to Java and it has matured into a serious contender as an enterprise development language [30].

Avancini suggested the structure for a Framework for Multi-Agent Systems (FraMAS) in 2000 [11] using a composition model and the *Decorator* pattern [121]. He presented the need to perceive, deliberate and communicate, but failed to highlight the OODA association required to embody cognitive functions. He did however espouse the need to abstract the problem from the supporting environment, making the agents easier to access the environment and its sub-systems.

Early examples using MAS involved concepts surrounding logistics operations like transportation and warehousing (including storage and retrieval activities). These require the ability to sort and make decisions to resolve conflicts, although many (like SOAR) were originally *end means* [110].

## 9.8   What Is a PID

Before we can access any process in memory, you need to gain a handle to its physical location. The most commonly used method to access software libraries is to link them during compilation. Debuggers then use symbolic references to report on the state of processes when active. Alternatively, you could determine the location of the process dynamically. When threaded, an agent is accessed in a sub-process (generally attached to the current JVM).

A Process IDentifier (PID) is a reference to a pipeline of a running process within the operating systems memory space. A programmer may need to know the PID of a running application, especially if that process must cooperates with other applications.

Presently Java is required to use either a Java Native Interface (JNI) procedure or wrap an Operating System (OS) script (.sh or .bat) which can be really inefficient. At least three requests have been made to provide an inbuilt method. These suggested methods include:

Modifying Java:    Bug ID 4250622

- public int getID() method in java.lang.Process; and
- hashcode() of java.lang.Process.

Enhanced Redirection:  Bug ID 4244896

- getID();
- getEnvVar(); and
- killProcess(String pid) From the java.lang.System class.

Modifying Linux:  Bug ID 4890847

- waitFor(long) - redirection to the native OS; and
- PID()

Listing 9.1 shows how to redirect the call through the OS. Even though Java threads use the native OS threads whenever possible, when using Linux, there is no mapping between Java threads and Linux threads. It may or may not be the parent PID.

Dedicated applications like the Java virtual machine Process Status tool (JPS) or a *JavaSysMon* can return the pid of the classname (see Listing 9.2).

```java
import java.io.IOException;

public class Pid {
    public static void main(String[] args) throws
        IOException {
        byte[] myData = new byte[100];
        String[] cmd = {"bash", "-c", "echo $PPID"};
        Process p = Runtime.getRuntime().exec(cmd);

        p.getInputStream().read(myData);
        System.out.println(new String(myData));
    }
}
```

**Listing 9.1** Bash file method

```java
String jps = [JDK HOME] + "\\bin\\jps.exe";
Process p = Runtime.getRuntime().exec(jps);
```

**Listing 9.2** Call to JPS

## 9.9  Parent Class

It turns out that after much research and many hours of toil, a reliable solution for obtaining a PID dynamically is possible. The management class provides the abstraction required in the JDK. It is possible to use the *RuntimeMXBean*, which has been available since Java SE5, to return the name of the running JVM. This is in the format (*processID@machineID*) for Windows, Linux and Mac OS X. Listing 9.3 shows a listing of how to use these classes to automate the task.

## 9.10   Dynamically Installing Agents

As discussed previously, the *instrumentation* package was introduced in Java SE5 and improved by including the *Attach* API in JDK SE6. This provides developers with the ability to *dynamically attach agents* to a running process at *run-time*, even if it resides in another JVM[12]. Unfortunately using command line arguments requires a number of fragile parameters to successfully invoke the agent.

Although this can be an error prone process, it can be automated and creates the ability of being able to re-transform classes. It is also possible to split the concept of using an *agent factory* into two parts. One part that assembles the object and the second part that packages and moves the capability. This can be achieved by applying reflective properties to the object to tag it with a label by calling the (**class.forName(the agent.name, true, classLoader.getSystemClassLoader**()) when loading the agent. Lines 10-14 strips out the PID which is returned at line 16.

```java
import java.lang.management.ManagementFactory;
import java.lang.management.RuntimeMXBean;

public class Parent {
  public static String getPID() {
    RuntimeMXBean mx = ManagementFactory.getRuntimeMXBean();
    String name = mx.getName();
    String namePID = "";

    int x = name.indexOf("@");
    if (x >= 0) {
      namePID = name.substring(0, x);
    } else {
      namePID = "0";
    }
      return namePID;
  }
}
```

**Listing 9.3** Parent Class

Developers should be wary of the need to provide the logic to create the agent and its processes. Java merely provides the ability to intercept the loader, enabling the modification of the original agents before it is instantiated. Thus the byte code can be amended at run-time using a call to **premain(String args, Instrumentation inst)**. This provides the indirection required to preserve the *single responsibility principle* of OOPL. The central difference being that developers are no longer forced to create

---

[12] The term agent that is used here can be a little misleading. It really provides the ability to load a package that can be run remotely or as a standalone framework. This may or may not include decision making and may or may not be in the same memory space or JVM. Either way it will enable the developer to attach a separate process to one that is presumed to maintain the user environment.

a class using a *main( )* method. This new class may now be run as an attachment using another classes *main( )* method to invoke or instantiate both classes. The calling class may also pass arguments to the attached class using a *String* that could be defined statically or dynamically. This would facilitate greater reuse or adaptability at run-time. Proxies are another form of indirection used to access objects remotely.

## 9.11   Proxy Classes

A proxy defines a protocol that enables communication between two objects. This protocol defines or wraps any number of *Constants, method signatures* or *Nested Types* required to conduct an exchange of information/state. All classes wrapped by an interface definition must contain definitions for all methods in the interface[13]. The types defined by the interface can be used anywhere it has scope, just like any other type (The interface and methods need to be public unless collectively packaged). Like classes, an interface can be extended by other interfaces. It can be used to upgrade a classes behaviour (upon exit) without the need for anyone to modify the original code. The syntax of implementing an interface can be quiet daunting and should be hidden in a package to simplify maintenance or reuse.

## 9.12   Dynamic Proxies

Dynamic proxies were introduced in Java SE1.3. It implements a list of interfaces specified at run-time when classes are invoked. It can be used to predefine a series of events to access the state in one or more classes. Typical examples show how to dynamically generate *get, set* and *isa* methods. The underlying type is managed using the handle provided by its proxy interface. Multiple reference can be generated for the same type. For instance, a *person* record could be accessed as an *employee, manager* or even a *customer*. The redirection provides the virtual reference to the same source using *java.lang.reflect.InvocationHandler*. Since JDK 5.0, RMI uses dynamic proxies to replace the need to manually generate stubs. The resulting capability now provides user friendly RMI that is dynamically applied at run-time. This functionality forms the core mechanism in many future autonomous and distributed functionality.

## 9.13   Agent Factories

Any piece of code is composed to exhibit a specified behaviour, manage data and/or state. An agent that operates across a distributed system generally adds mobility to this concept. Given that the JVM provides an homogeneous environment for

---

[13] In Java a class is wrapped with an interface using the implements key word in the same manner that the extends keyword is used for inheritance.

execution, it protects the developer from the nuances of OS complexities. Although it is necessary for agents to migrate to the source[14], not all Agents need mobility, although they may exhibit the same capability to complete task both onboard or off board. The *agent factory* needs to be able to generate agents (or preferably teams) with a specified capability that can be deployed, if required. This can be achieved in two parts, one for the capability and another for *tagging, logging, exchanging/communicating* or *mobility*. The secondary protocols can extend the capability using whta looks like *polymorphism* at run-time. In Java this is achieved by *implementing* a wrapper. The wrapper can be attached manually at compile-time by the designer or dynamically during run-time using *dynamic proxies*. Given the exchange protocol can be tailored to a specified IDL, it is possible to exchange information with another agent written in another language. It would also be possible to ascertain the *remote* host language and when available, package a language specific agent to migrate and execute in a language dependent environment using the same IDL. To achieve many of these tasks, developers typically need to understand many protocols, services and APIs. Java provides package that enable developers to implement; SOAP, REST, RMI, XML/XOM, WSDL and many other techniques that enable agency in world class applications.

## 9.14  Tagging Interface

An interface in Java enables the programmer to abstract a new reference type. It has no code, but can be *implemented* as a direct extension of one or more interfaces. Thus allowing multiple objects to *extend* (appear to inherit) one or more common behavior(s) as shown in Listing 9.4. The tagging interface would provide the inner most abstraction and each class must provide its own implementation of *tagAgent* (see Listing 9.5). This would seamlessly provide the functionality required to automatically label the appropriate amount of reflective tags required by the system to identify the agent being wrapped. For instance adding a name based on the environment, source and issuing a unique ID (normally *Serialized*) to assist in managing entities in that system. The interface defines the abstraction where the implementation (as the name suggests) provides the process.

```
1  public interface AgentTaggingInterface {
2
3      // abstracted tag member
4      public int tagAgent(Agent team);
5  }
```

**Listing 9.4** Agent Tagging Interface

---

[14] Local processing reduces the network bandwidth, as it only returns the result.

```
1  public class AgentTeam implements AgentTaggingInterface {
2      public double id = 0;
3
4      // constructors
5      public constructor1 () {
6          doSomething ();
7      }
8
9      // implement AgentTaggingInterface
10     public void tagAgent (Agent team) {
11         addID ();
12     }
13 }
```

**Listing 9.5** Agent Tagging Implementation

```
1  public class LoggingHandler implements InvocationHandler {
2
3      protected Object delegate;
4
5      public LoggingHandler (Object delegate) {
6          this.delegate = delegate;
7      }
8
9      public Object invoke (Object proxy, Method method,
           Object [] args)
10         throws Throwable {
11
12         try {
13
14             System.out.println ("Method called via "
15                 + method + " @ "
16                 + System.currentTimeMillis ());
17             Object result = method.invoke (delegate, args);
18             return result;
19         } catch (InvocationTargetException e) {
20             throw e.getTargetException ();
21         } finally {
22
23             System.out.println ("Call Logged by "
24                 + method + " @ "
25                 + System.currentTimeMillis ());
26         }
27
28     }
29
30 }
```

**Listing 9.6** Logging Interface Implementation

## 9.15    Logging Interface

To organise any form of indirection, the programmer must invoke a handler to acquire access. Logging relies on reflection to provide human readable references, hence the need to publicly tag classes (agents). Listing 9.6 shows a method that prints a simple time stamp when called. There are a host of examples and tutorials available to assist readers who wish to pursue this form of interface. Many tutorials call proxies, *decorator* to existing classes. Early implementations used RMI, EJB, and JAX-RPC. More recently, the trend is to use the *InvocationHandler interface* to create dynamic proxies.

## 9.16    Communication Interface

The communication interface provides the hooks required for an agents to exchange information. Typical hooks may include *requestToSend( ), beginTransaction( ), transferResources( ), beginTransfer( )* and *endTransfer( )*. There are also a number of agent communication languages in common use. These include; ACL, KQML, FIPA and FIPA/ACL. Although this was a topic of research conducted at KES, a multi-protocol Agent becomes unwieldily and at the time, was very difficult to achieve interoperability without knowing the destination protocol. No further effort has been expended on this project, however it would be far easier given that those protocols carried some form of identifying *meta-data*. Platform independence, Language independence (both for execution and communications/sharing) and the use of ubiquitous Protocols is a visionary goal that can only be made possible using self discovery. Therefore reflection and persistence become central to the implementation of future applications.

## 9.17    Mobile Interface

To achieve mobility an agent must be transported from the source to the target (client/server), complete the desired task and return with an appropriate response. It must also contain the ability to autonomously package and transport any code/state required. More complex implementations, may require a series of transactions based on external events. If that event is periodic in nature or the network connection is congested or unstable, that complexity escalates. The typical move transaction is shown in Figure 9.1. The code required has been discussed in many publications and forums. It has now been integrated into a series of protocols, such as FIPA.

A number of Enterprise applications have mobility built into their middleware products. If you are interested in finding out more, investigate the Mobile Agent Facility (MAF), Mobile Agent System Interoperability Facility (MASIF) and Object Management Group (OMG) projects.

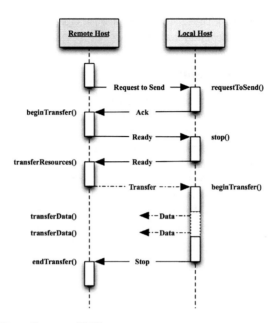

**Fig. 9.1** Move Event Sequence [355]

## 9.18  Other Interfaces

The logging interface can be used to seamlessly automatically wrapping an agent
with the appropriate code required by the host system to log the actions of this agent.
Additional interfaces can be used as wrappers to progressively add more automation,
such as data exchange, communication and mobility. Alternatively, an interface can
be extended by one or more interface types as shown in Listing 9.7. Such abstraction
is convenient, but not without issue.

```
1 public interface MyInterface extends Interface1 ,
    Interface2 {
2    // constant declarations
3    int i = 15;
4    double d = 1.2345;  // variables
5
6    // method signatures
7    void doSomething (int i, double d);
8    int doSomethingElse(String s);
9 }
```

**Listing 9.7** Using Multiple Interfaces

The biggest difficulty with abstraction is the need to reproduce the implementation of all abstracted method declarations in each and every extension.

## 9.19   Agent Team Package

Internally the class structure can provide a team hierarchy from system to supervisor, team, group, capability, right down to dynamically accessible functionality. Using AI, environmental stimuli can be pre-processed to drive context selectable knowledge-bases that can be used to dynamically modify strategy, roles and/or entity behavior.

Figure 9.2 shows the basic concept of a agent team (as separate classes), where any class can be implemented using one or more wrappers (recursively) to enhance its run-time capabilities. An example could include the need to modify the method of exchanging data or communicating with one or more classes when instantiated[15]. In a game environment it may be necessary to have information about the environment updated statically (periodically), while the mission critical parameters (like health) may need to be updated dynamically (as they change). The structure and composition of the team could be constructed statically or even dynamically using the appropriate tools or libraries.

## 9.20   Agent Factory Design

An agent factory demonstrator is an application written to experiment with the concept of providing dynamic functionality within an agent team environment. The basic concept used an IDE coupled to an agent factory that produces a pool of agents that can dynamically alter its functionality during run-time. The pattern for the factory agent is provided in Figure 9.3. These agents are instantiated with a number of states used to represent the state of the model at any point in time. The state of agents would change throughout their lifetime depending on the state and conditions they are at or the work they are doing. Some states are common among all agents while other states are particularly for certain agents. For example, some common states of agents include *Initiated, Active, Suspended, Uninitiated, Unknown*. All agents start with initiated state and their state becomes active when they are ready to do their job. Any agent can also be suspended. This would mean its thread would be suspended until it is reactivated. When the agent's thread is destroyed, its state is set to disposable or even to an unknown state. As mentioned before, some of these states are defined within selected agents. The state can be published and accessed by other entities. For example, client agents can reflect its state, such as *searching*, when they are searching for an expert agent, or *waiting* when they are waiting for resources or users to respond. Another state of client agents is defined as *Task_is_done* or *complete*, which shows that they have accomplished their mission and declare that they are available to engage in a new mission. They can be switched between *halt* and *wake, sleep* or even *pause* and *standby* during the process of achieving a goal. The agents current

---

[15] The dotted lines are included to indicate the example is not complete and that other classes or interfaces could be displayed or incorporated as required.

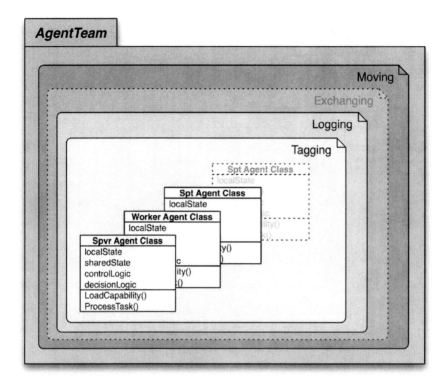

**Fig. 9.2** Concept of the Agent Team Package

state will depend on the scenario and goal being processed. This can be used to help in achieving a more sophisticated or intelligent communication model among agents in a team. For example, the expert agents can indicate that they are *Busy* or they are doing other jobs. This means that other agents have to wait or seek an alternative expert within the system.

## 9.21  Agent Factory

There are several groups who have attempted to produce *agent factories* in the public domain. Some have provided little or no documentation, however a growing number of open source projects are attempting to remedy that problem. The Agent Factory Standard Edition (AFSE) is one of these projects. This is FIPA compliant, modular and extensible. Earlier efforts include: Agents Channelling ContExt Sensitive Services (ACCESS)[16], Clarity[17] and Agent Factory Agent Programming Language (AFAPL)[18].

---

[16] See www.agentfactory.com/index.php/The_ACCESS_Architecture.

[17] See http://www.clarity-centre.org/.

[18] See http://www.agentfactory.com/index.php/AFAPL2.

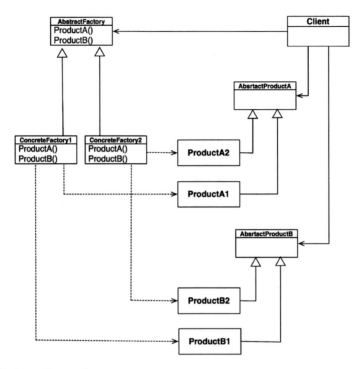

**Fig. 9.3** Agent Factory Pattern

**Table 9.4** Comparison of Migration Model, Environment and Platform [129]

| Migration Model | Environment | Platform |
|---|---|---|
| Homogeneous | Either both *virtual* or both *physical* | Both source and target identical |
| Cross-Platform | Either both *virtual* or both *physical* | Divergent source and target |
| Regenration | One *virtual* and one *physical* | Both source and target identical |
| Heterogeneous | Both different | Both different |

There are a variety of scenarios that programmers could encounter when they choose to migrate agents. Factors that need to be considered include the platform, its operating system and the environment the agent would encounter. Several other factors include: *endian*, architecture, microprocessor, communication medium, environment and reliability. Table 9.4 provides a quick summary of the relationship between published migration models, possible environments and platforms. We only consider an aggregated collection of agent-based functionality using a generative model. We also confine our discussion to the Java development environment using the JVM.

Languages like PROcedural LOGic (PROLOG) and LISP rely on sequences of procedural statements to progressively compound or expand an agents capability (ideally dynamically). Using this extensive approach the functionality can adapt in an unstable environment. A fairly coarse script or primitive language based on an agent capability could easily be implemented to achieve sophisticated or complex tasks using a mixture of predefine structure and business logic. The same concept can be extended to our agent factory. Using a standard IDL, structured ontology and uniform communications policies, flexible capabilities can be generated or even regenerated (weak or strong migration). As indicated above, we concentrate on one language and will also leave discussion about regenerative models to researchers currently developing this area.

Both models require a *schema* or *blueprint* in order to collect and assemble a series of resources required to achieve the desired goal. To generate or manually assemble a specific capability, a programmer would generally:

- identify the schema required to service the goal,
- obtain the resources and identify the IDL required to achieve interoperability,
- determine the ontology that best supports the capability, and
- identify the operational configuration and rational.

Given the capability is created, it must either be wrapped with the required interoperability functionality. The alternative is for the programmer to tackle the complexities required to embed that functionality. A number of these wrappers exist within several frameworks. Examples include: FIPA-OS, JADE and AGENTSCAPE. As these systems evolve, based on existing standards and patterns, the ability to reuse these components increases. With the increased availability of components, libraries can be developed and accessed to assemble capabilities composed as MAS. Scripts required to automate this process have proven successful in enterprise level applications based on a dedicated blueprints. This topic is complex, which is the main reason agent factories are being developed. Technology continues to evolve, making it possible to have decision making logic, dynamically generate a given capability using a specified *blueprint*. That blueprint could be customised dynamically, based on specified environmental parameters and aid decision-making without the need for complex and costly human intervention.

## 9.22  Why Agents

As stated previously, agents are increasing being used to solve progressively more complex problems. As we approach applications that solve real-world problems, the skills required have risen dramatically. Practical examples of where agents are currently used, include: spell checking, spam filters, travel and event booking systems.

The code required to create an agent factory will focus on the needs of the programmer. Code could be linked at compile-time, but more preferably, instantiated and attached during run-time. In order to abstract the inherent complexity of this task, a factory is required. It needs functionality that dynamically wraps and packages a given capability. Intelligent systems can be developed to modify existing code, even when in memory, without loss of state or downtime. Next we discuss the problems related to providing the strategies and logic required to solve a sudoku puzzle. We then provide a practical case study and exploit some of the techniques discussed in Chapter 11.

*"AI's record of barefaced public deception is unparalleled in annals of academic study [386]."*

J. Vaux and R. Dale

# 10

# Case Study Background - Sudoku

Many algorithms exist to solve all sorts of puzzle. A lot of these have been implemented in software. The reader is invited to seek out alternative solutions and are challenged to provide their own implementation(s). Russell and Norvig [320] provide the necessary background to the history and adaptation of many of these techniques within the AI domain. After introducing the problem, we will briefly discuss Heuristic search [16][1], Brute Force (exhaustive) and Backtracking [401]. The only mandatory condition for a Sudoku puzzle is that it must have a solution, although it should be solved logically and without guessing or search.

## 10.1  What Is Sudoku

Given the name of this game, many people believe it to be Japanese, however there are many claims to its origin[2]. There is some evidence that a similar puzzle appeared in a French newspaper in 1895, the current '9x9' form of the puzzle[3] was published in newspapers in New York[4] during the 1970's, migrated to Japan[5], then England [28][6] and back to New York in May 2005 [23, 83, 148, 341]. We make no claims about

---

[1] Generally using Ends Means Analysis (EMA) algorithms.

[2] Possibly as early as 1783 as a mathematical model by Leonhard Euler, called *Latin Squares*.

[3] A puzzle is generally symmetrical given an '$r*c$' construction. Where $r$ normally equals $c$, however a number of variations have started to appear.

[4] Dell Pencil Puzzles and Word Games - Issue 16 [290].

[5] Popularised by Nikoli - see http://www.nikoli.co.jp/en/

[6] After Wayne Gould developed a computer program to produce puzzles quickly (see his web-site www.sudoku.com), it was used to promote Sudoku in the *Times* in Britain, starting on 12 November 2004. These also appeared in the *Conway Daily Sun* (New Hampshire) in 2004.

J.W. Tweedale & L.C. Jain: Embedded Automation in Human-Agent Environment, ALO 10, pp. 145–162.
springerlink.com                                                    © Springer-Verlag Berlin Heidelberg 2011

its geneology, terminology or even the strategies presented, however we do refer to adequate sources of information (both on-line and in print). There are many publications with a vast array of puzzles with varying levels of difficulty. Search Amazon for Wayne Gould or surf to[7] for some alternatives[8]. These references are intended to aid interested readers in forming their own opinions. Our aim is merely to use the puzzle concept to illustrate several examples of AI techniques using agency theory.

Many websites promote this strategy based puzzle. Several include; Sudoku Dragon[9], Solving Sudoku[10] and Sudoku Essentials[11]. Most provide detailed descriptions on strategies to solve the puzzles and a variety of examples.

Sudoku represents a three-dimensional puzzle that requires the reader to complete the puzzle using a series of numbers, letter, symbols or colors (generally '1-9'). It only has approximately 5,500 million workable solutions (After you account for symmetries; such as rotation, reflection, permutation and relabelling)[12]. Most variants of Sudoku puzzles are created without a symmetrical constraint using no fewer than 17 values (This still results in over 48,000 unique examples)[14]. This would fill an entire year, if you could stay awake long enough and complete every puzzle in under 10 minutes (otherwise 131 years at one per day). To understand Sudoku, you should become familiar with a variety of terms or descriptions (These are sourced from many disciplines).

### 10.1.1   Terminology

The Sudoku terms being used in this chapter include: Cell, Row, Column and Blocks. Any constraints given are called clues and possible values are generally labelled as candidates. A number of interchangeable alternatives that have been encountered include:

| | |
|---|---|
| Cell: | Square, Box, Element; |
| Row/Column: | Group, Chain, Vertex; and |
| Block: | Grid, Zone or Region. |

The first cell of a puzzle is shown in Figure 10.1. A *cell* is described as a *square* and displayed for the user to enter a valid selection from a predefined range of numbers, letters, symbols or colors. These are placed horizontally to form a chain of boxes into a row. Separate rows are stacked vertically to form a matrix of cells which can be addressed using a row/column coordination system.

---

[7] http://sudokusolver.com/books/

[8] See http://mapleta.maths.uwa.edu.au/~gordon/sudokumin.php

[9] See http://www.sudokudragon.com/sudoku.htm

[10] See http://www.angusj.com/sudoku/hints.php

[11] See http://www.sudokuessentials.com/

[12] Felgenhauer and Frazer Jarvis[13] in 2005 is roughly $1.2*10^6$ times the number of Latin square possible.

[14] See http://www.csse.uwa.edu.au/~gordon/sudokumin.php

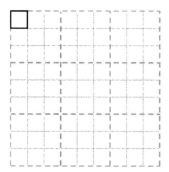

**Fig. 10.1** Sudoku Square or Cell

The images shown in Figure 10.2 display how nine squares can be assembled to represent a row, column or block. The placement of numbers, letters, symbols or colors within any cell can be represented as a series of numbers (permutation set with respect to the range and combinations formed).

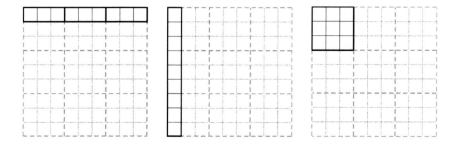

**Fig. 10.2** Sudoku Row/Column/Block Representation

Figure 10.3 displays a graphical representation of the number of paths that can be determined for each of the nine possible candidates relating to a '9*9' puzzle (which is '3*3' Blocks)[15]. Therefore a puzzle consisting of '9*9' matrix of cells is considered to contain nine sub-groupings or Blocks of '3*3' cells, each square containing a constrained series of numbers '1' through '9'.

Every row, column or block consists of a collection of the overall matrix of squares representing the puzzle. Nine cells in any row, column of block, must contain $m^4$ cells. Each cell holds only one of the $m^2$ values for any series (row, column or block). To

---

[15] Please note, the term Block and Puzzle are not interchangeable, however many people use the alternative terminology as they use different methods to solve a puzzle. In a symmetrical Sudoku, '$m*n$' Blocks becomes a '$m*n$' x '$m*n$' puzzle (when $m=n$, this becomes an $m^4$ cell array. Alternatively you could write an puzzle with '$m*m$' Blocks or an $m^2 * m^2$ puzzle, again where $m=n$.

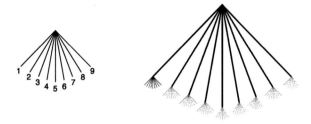

**Fig. 10.3** Simple Permutation Graph or Tree (with/without leaves)

the right of Figure 10.3 is the tree with leaves, which represents the possible paths for the entire 81 constraints represented by a '9*9' puzzle series. This tree can be further expanded to represent the full compliment of available series in each row, column or block. This explosion of alternatives is not practical to display given there are 27 additional constraints[16].

The tree shows how the relationship of the constraints and how a path can be planned. AI has many methods that can be used to traverse a tree in seeking a solution (some more efficient than others) [338].

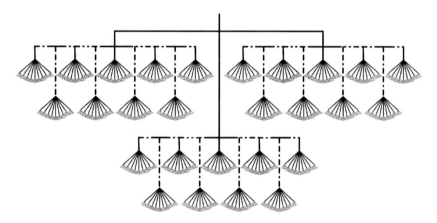

**Fig. 10.4** Tree with a Trunk containing limbs, Branches and Leaves

---

[16] The Tree shown in Figure 10.4 has nine branches, each containing nine leaves (81 in this example). This is compounded when the tree is redrawn at the trunk, with 27 limbs, each attached to nine branches, with nine individual leaves (2,187 in total). If you discount the first branch, 1944 constraints remain to be explored (pruned). You can further optimize the problem by removing the symmetrical similarities, leaving 972 constraints to deal with [159].

## 10.1.2  Representation

Puzzles are available in many forms and variation as are its contents. The most common representation of a Sudoku puzzles lies in the form of numeric matrices. These are grouped as an array of Blocks '*m*' rows wide by '*n*' columns high as shown at the top left of Figure 10.5. Each Block contains an array of cells consisting of '*m*' rows by '*n*' columns which is depicted in a matrix of '*m* * *n*' x '*m* * *n*' squares. The overall puzzle is addressed using the coordinate system shown at the bottom of Figure 10.5. This can be row/column or more accurately block/row/column. To reduce all confusion, we will use letters to represent rows and number to represent columns (Note that 'I' is skipped to ensure it can not be confused with number '1'). Our puzzles flow from left to right, then top to bottom. The boxes are constrained to a unique series of numbers, letters, colors, symbols or even pictures. From this point, that will be the numbers '1-9'. Other forms of the puzzle do exist, but will not be subject to

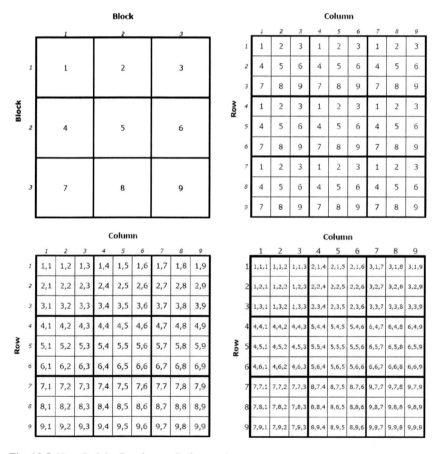

**Fig. 10.5** How Sudoku Puzzles are Referenced

discussion here. Some are based on the *hypercube* (such as *Samunamupure*), rectangular blocks and other shape enumerated results [23]. Examples include:

- 3doku Stertenbrink[17],
- Disjoint Groups Russell[16],
- Hypercube Stertenbrink[16],
- Magic Sudoku Stertenbrink[18],
- Sudoku X Russell[16], and
- NRC Sudoku Brouwer[19].

A Sudoku puzzle normally contains some initial candidates. The number of clues and their location relates to the level of difficulty. Most puzzles are classified into categories. Generally these will be easy (beginner or novice), mild (challenging) or hard (difficult approaching diabolical). Figure 10.6 provides an example of how a '9*9' Sudoku puzzle, with 23 clues, is displayed. In this case the example shows the 'AI Sudoku' which is characterised as the most difficult puzzle in circulation [153].

| 1 |   |   |   |   | 7 |   | 9 |   |
|---|---|---|---|---|---|---|---|---|
|   | 3 |   |   | 2 |   |   |   | 8 |
|   |   | 9 | 6 |   |   | 5 |   |   |
|   |   | 5 | 3 |   |   | 9 |   |   |
|   | 1 |   |   | 8 |   |   |   | 2 |
| 6 |   |   |   |   | 4 |   |   |   |
| 3 |   |   |   |   |   |   | 1 |   |
|   | 4 |   |   |   |   |   |   | 7 |
|   |   | 7 |   |   |   | 3 |   |   |

**Fig. 10.6** Example Puzzle - 'AI Escargot' [153]

### 10.1.3  Mathematical Analysis

Puzzle solving in Sudoku falls into two main areas: Grid analysis and Puzzle analysis. These predominantly use combination or permutation group theory. Research

---

[17] http://www.sudoku.com/boards/viewtopic.php

[18] http://www.setbb.com/phpbb/viewtopic.php

[19] http://homepages.cwi.nl/~aeb/games/sudoku/nrc.html

has shown that there are significantly fewer valid Sudoku solution blocks than Latin squares. This is due mainly because of the additional regional constraint imposed. Many techniques developed to analyze Latin squares still apply. You could use Cayley tables to calculate the size and composition of a valid Sudoku puzzle combinations or facilitate pattern matching. Alternatively you could discount the symmetrical nature of solutions using Burnside's lemma [42] to translate from one valid block to another, based on:

- Relabeling symbols (9!),
- Band permutations (3!),
- Row permutations within a band (3!3),
- Stack permutations (3!),
- Column permutations within a stack (3!3), and
- Reflection, transposition and rotation.

## 10.2   Puzzle Theory

Sudoku is all about permutations, with an extra twist of logic. The number of permutations found is calculated using the Factorial of '$m * n$'. There are 362,880 possible ways of ordering a '$3*3$' set of symbols in a puzzle row, column or block. In a '$9*9$' puzzle, using the candidates '1-9', every row, column and block must sum to equal 45. Although it is not practical to use the associative rule, the order in which you add these numbers together does not matter. Again you could multiply each number to derive a total, however cannot subtract or divide indiscriminately because you get different results. To obtain a unique way to determine missing candidates you could substitute each value with its *primorial* equivalent [46][20]. Once you multiply the primes, you can subtract this from ordinal present value, and repetitively deduce missing number(s) [59]. For example, given a series '1, 4, 7, 8, 5, 2, 9 & 3' we can determine '6' is missing. The primordial for 9! is 223,092,870. Using *Group Theory*[21] you divide the known values successively by the existing value prime substitutes and you get 13 (or 6 from Table 10.1).

## 10.3   Rules

There are only three rules to Sudoku.

1. Each cell in every row must contain only one entry, however that row must contain one of every entry within the available range.
2. Each cell in every column must contain only one entry, however that column must contain one of every entry within the available range.
3. Each cell in every block must contain only one entry, however that block must contain one of every entry within the available range.

---

[20] *Godel* numbers are created when you substitute predetermined, unique free variables with all known occurrences of every number in a series.

[21] This concept is also known as *ring* theory.

**Table 10.1** Table of Prime Equivalents

| Number | Prime |
|--------|-------|
| 1 | 2 |
| 2 | 3 |
| 3 | 5 |
| 4 | 7 |
| 5 | 11 |
| 6 | 13 |
| 7 | 17 |
| 8 | 19 |
| 9 | 23 |

## 10.4 Strategies

We do not believe it is necessary to provide a tutorial or other forms of instruction about Sudoku, or the myriad of strategies that have evolved or been applied to Sudoku. The following list includes some of those encountered. A description has been provided in most cases. If you require more information, the reader needs to only search on the web or read more widely (see Mailer [236], Crook [69] or Mepham [251]). Some of the broader approaches include:

Looking past the linear dimension: Each square is a member of three regions (row; column and block). This provides a three dimensional array of possibilities as constraints apply equally to all groups. So if a column presents with a specific sequence, those constraints can be applied to all of the associated row, column or block.

Applying the rule of K: You can sum of any '9*9' Sudoku series in any order and obtain a total value of 45. Rather than using a list, you could use a sum calculator to reduce the number of alternatives remaining.

Using the general permutation rule: If you can identify a group within a permutation series that is restricted to the same number of squares then you have a Sudoku permutation rule. This applies to twins, triplets and quintuplets in terms of the size and the sub-group you can spot elsewhere.

Most strategies are generally categorized by level, ranging from basic to advanced standing. A number of those encountered are listed by category:

### Basic Strategies

The One Value: All cells must contain a value within the puzzle range, however there must not be any duplicate values within any single row, column or block. For instance a '9*9' puzzle uses all the numbers '1' through '9'.

The Single Possibility: This rule signifies that a cell can only be the remaining candidate. An example, based on the series '1,2,3,4,5,b,7,8,9' shows the missing value '6' must be assigned in the appropriate blank cell (where 'b' is the blank).

Remaining or Only choice: Similar to the Single Possibility Rule. Given all but one of the possible value, there may be only one possible choice remaining. This is generally applied iteratively, each time a cell is solved.

Only cell or only square rule: It is often the case that an intersecting (or shared) group will mean that a number can not go in one of two squares. You are left with an 'only square' for a candidate to be placed.

Hidden Single: The rule assumes the player maintains a list when attempting to eliminate or exclude possible values. After evaluating the series '1,b,b,4,5,b,7,8,b' for a given row, column or block, a number of common possibilities may occur (due to intersecting constraints), for all but one lone cell. Thus the lone cell can be assigned the isolated value.

Scanning: Candidates can be eliminated due to constraints imposed through associations within others groups. Take the *AI Escargot* example in Figure 10.6. By highlighting the '1s' in the two first two columns of the left most column of blocks. Due to Block one and two, column three of lowest block must contain a '1'. Due to row seven, it can only go in row eight (The coordinate of this cell would be 7-8-3 using the block, row, column system or simply H3 using the row/column system.)[22]

Elimination: This is similar to Scanning but extended to the whole puzzle. By focusing only on one number at a time it is possible to constrain the location of values to a sub-group row or column. Given a number of existing clues it is possible to determine it to be the only value for certain blocks. Again look at only the '9s' in the AI Escargot example. By focusing only on the '9s' they can immediately be constrained to block two at B4 and B6. Looking down from the right most block, it can also be constrained in block nine at G9 and J9. Unfortunately there are no other nines to work with and no other clues available to help eliminate these values at this point. If a nine existed elsewhere (say , it could be eliminated in further locations. Unfortunately another rule is required.

---

[22] Some designers use a spreadsheet style coordination system. This is achieved by replacing each row with the equivalent letter. Therefore 7-8-3 becomes 7-H-3 or H3 for simplicity. The later is preferred because you can use this notation with values, such as H3-1.

Shotgun:            This is a form of random trial and error, followed by scanning
                    with elimination. Without a systematic approach, this will
                    rarely solve the puzzle.

## Intermediate Strategies

Locked Candidate:   This is where other candidates can be excluded from in-line
                    blocks due to the fact that a specified candidate must occur
                    in a constrained position within puzzle. If you assess the '1s'
                    for blocks 1, 4 and 7 in Figure 10.6. We know that '1' must
                    occur in H4. The result is the '1s' in block 8 are constrained
                    to J5, J6 and J7 and eliminated from the rest of block 8. If
                    there was no '1' present in block 9 it would also be eliminated
                    from row J.
Single Possibility: A 'single possibility' is represented in a series within a closed
                    loop of symbols. The way that groups intersect with each
                    other may indicate that there is only one unallocated cell.
                    Therefore there remains only one possible candidate. Alter-
                    natively, one block has two naked twins in a single row. Us-
                    ing the twin rule, an intersecting value can not occur in the
                    remaining unallocated cells, leaving the alternate to be lo-
                    cated in the remaining square.
Naked Twin:         A 'naked twin' represents two candidates with only two pos-
                    sible locations. These can safely be used to restrict or deduce
                    where a specified value can go. By using the knowledge that
                    a symbol may occur only within one subset, you can also
                    determine that it cannot go elsewhere in that series. A naked
                    twins appears when a pair of numbers are presented or re-
                    main within a series, that is, both have been excluded within
                    other possibilities in that series.

## Advanced Strategies

Hidden Twins:       The twin (or triplet) exclusion rule relies on people spotting
                    matching patterns hidden within other possibilities within a
                    series (row, column or block). When the identified pair can
                    only go in the matching two squares, then those numbers can
                    be excluded from other blank cells within that group.
X-Wing:             The X-Wing[23] represents four groups of logically inter-
                    linked pairs that form a 'box'. This concept relies on the
                    premise that a resolution must exist in one of the two di-
                    agonals. As these numbers must be unique in either row
                    or column, all numbers outside that formation can there-
                    fore be safely eliminated to help solve other cells within the

---

[23] The term X-Wing is probably derived from the name of Star Wars fighter.

|              | associated row or column. This strategy is generally limited to difficult puzzles. [24]. |
|--------------|--------------|
| X-Y Wing:    | Once you have removed all the naked, hidden and locked singles or twins, you need to try more complex solutions. The XY-Wing pattern is made up of three cells that form a 'Y'. Although you need to stretch your imagination some what to see a 'Y' pattern (more like two sides of an open box). Each cell contains only two numbers, however they represent three unique candidates between them (label 'X', 'Y' and 'Z'. All three of these values must appear in the cell representing the corner of the open box). The root or base of the 'y' contains 'XY' and each branch 'YZ' and 'XZ' respectively. By projecting the branch cells by row and column (to the open corner), you can safely remove the 'Z' element from that cell[25]. |
| Sword Fish:  | Swordfish is a further complication of X-Wing, which allows three linked pairs (one is an extension 'sword' jutting out on one side of the box formed by the four pairs. This formation is rare and examples are hard to find, however it can be used in difficult situations. |
| Finned Sword-Fish: | As discussed in Sword-Fish, not every cell (triples) in a 3 (rows) by 3 (columns) are required to contain the selected value. Generally you ignore any rows containing more than three possibilities, however where the one column only contains an extra value, the Finned Rule allows you to ignore it and confine the eliminations to the box formed. By focusing on the possible '9s' in the AI Escargot example, a 2-3-3 Finned Sword-Fish is formed, with G4, J4, G6 and J6 being eliminated. This still leaves 19 alternatives, making solving the remaining '9s' fairly difficult. In fact, in this case, trial and error is your only option (at least until I discover further strategies). |
| Jelly Fish:  | This approach is similar to Swordfish. You assess any four intersecting rows and columns that contain the same candidate. That candidate must exist up to four times in each row and lie in exactly four columns. When this occurs, the candidates in the remaining rows can be eliminated[26]. |
| Squirm Bag:  | Not much beats a Jellyfish, except a Squirmbag. This is based on the premise of matching a single candidate in five rows and columns. Although we found no examples of its use in |

---

[24] For an example, see http://www.angusj.com/sudoku/hints.php

[25] For a worked example, see
http://www.sudokuessentials.com/Sudoku-XY-Wing.html

[26] For a worked example, see
http://decabit.com/Sudoku/Techniques/Jellyfish.aspx

a '9*9' puzzle, it can be used for puzzles with larger dimensions[27].

Backtracking or Trial and Error: Guaranteed to work, but reserved for use only after you have logically determined solvable candidates. This involves the successive selection of possible values which are progressively tested in order until an error occurs. At this point the puzzle is unwound and an alternate selection is made.

## 10.5   Walk Through

Even puzzles with few clues can be solved rapidly with a few simple strategies. Examine the puzzle at the left of Figure 10.7. This has 17 clues with three clues at most in any one row or column. Even seven of the Blocks presented have two or less clues. So how difficult is it to solve manually. Follows is a walk through using only the Single Value rule. The process applied is to programatically check for any singles solutions, then isolate hidden singles until no further solutions appear. Being a computer program, not all suggestion will appear obvious to humans, however they do become clear within one or two moves. Initially you can assess the puzzle and create a list of all possible values in every cell (as display to the right in Figure 10.7. By reviewing the list, you can attempt to identify any hidden singles. Given you find one or more values, the process becomes iterative. Scanning from left to right, then top to bottom, you will discover a hidden single in column 'A' at row '5' (A5). This can be flagged as a '5'. The notation for this reads as 'A5' which becomes the value '5', based on that column 'c', or A5-5c[28], similarly A9 becomes a 1 by scanning its column. This becomes A9-1c [resulting in 2 solves]. Again revise your list and look for more hidden singles. This results in:

- B4-1r, B5-7c, G2-1r[29] and G7-4r (resulting in no new Singles) [4 solves],
- Again revise list, C2-5r, D1-1r[30], F3-5c, F6-1c and J8-1r&c (resulting in no new Singles) [5 solves],
- Again revise list, E8-5r [1 solve].

  - where H8 can now be determined to be '6' [1 solved];
  - then H8-6 enables H9 to become '8' [1 solved];
  - then H7 to '5' [1 solved];
  - with J7 becoming '2' completing this block; and
  - A flow on effect reveals that H6 can also be determined to be 9 [1+1 solved].

---

[27] For a worked example, visit 'Cluemaster' at
http://www.cluemaster.com/SolvingTechniques/JellyFish.asp

[28] A series is scanned within a row (r), column (c) or block (b). Clues can be revealed simultaneously in a row and column (r&c).

[29] This solution (and the next) isn't obvious to humans until after E8-5r.

[30] Again, this solution isn't obvious for several moves.

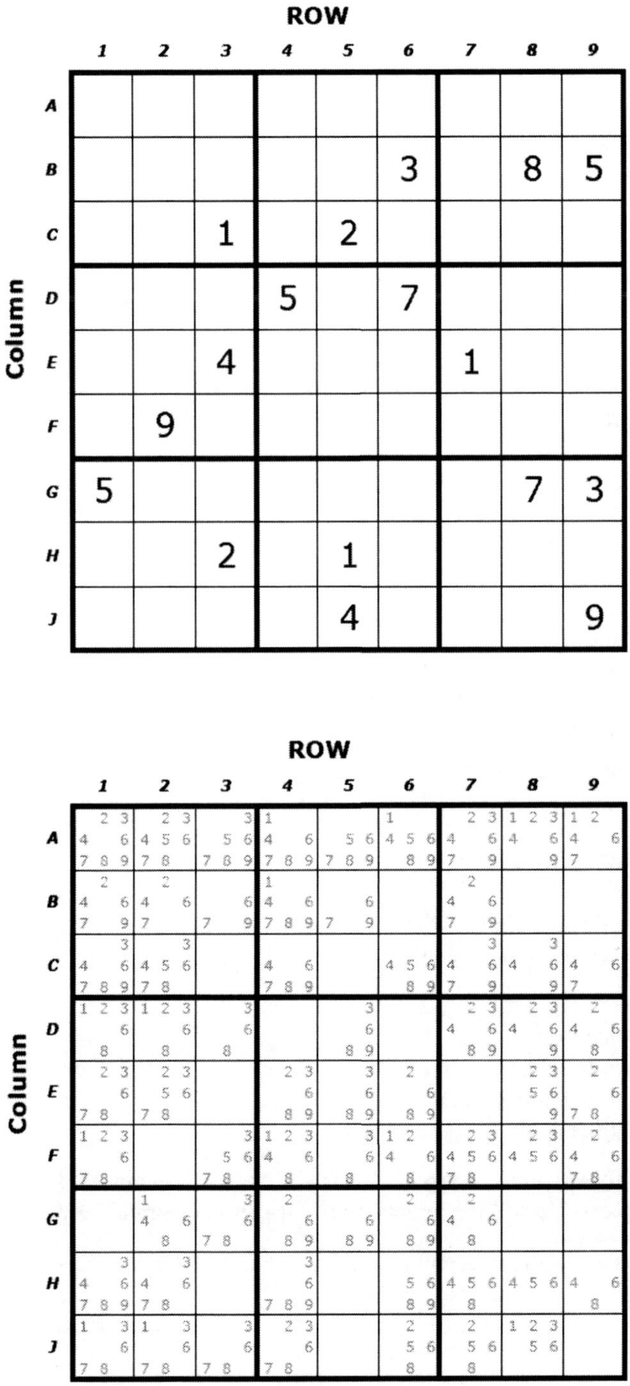

**ROW**

| Column | 1 | 2 | 3 | 4 | 5 | 6 | 7 | 8 | 9 |
|---|---|---|---|---|---|---|---|---|---|
| **A** |   |   |   |   |   |   |   |   |   |
| **B** |   |   |   |   |   | 3 |   | 8 | 5 |
| **C** |   |   | 1 |   | 2 |   |   |   |   |
| **D** |   |   |   | 5 |   | 7 |   |   |   |
| **E** |   |   | 4 |   |   |   | 1 |   |   |
| **F** |   | 9 |   |   |   |   |   |   |   |
| **G** | 5 |   |   |   |   |   |   | 7 | 3 |
| **H** |   |   | 2 |   | 1 |   |   |   |   |
| **J** |   |   |   |   | 4 |   |   |   | 9 |

**ROW**

| Column | 1 | 2 | 3 | 4 | 5 | 6 | 7 | 8 | 9 |
|---|---|---|---|---|---|---|---|---|---|
| **A** | 2,3,4,7,8,9 | 2,3,4,5,6,7,8 | 3,5,6,7,8,9 | 1,4,6,7,8,9 | 5,6,7,8,9 | 1,4,5,6,8,9 | 2,3,4,6,7,9 | 1,2,3,4,6,9 | 1,2,4,6,7 |
| **B** | 2,4,6,7,9 | 2,4,6,7 | 6,7,9 | 1,4,6,7,8,9 | 6,7,9 | 3 | 2,4,6,7,9 | 8 | 5 |
| **C** | 3,4,6,7,8,9 | 3,4,5,6,7,8 | 1 | 4,6,7,8,9 | 2 | 4,5,6,8,9 | 3,4,6,7,9 | 3,4,6,9 | 4,6,7 |
| **D** | 1,2,3,6,8 | 1,2,3,6,8 | 3,6,8 | 5 | 3,6,8,9 | 7 | 2,3,4,6,8,9 | 2,3,4,6,9 | 2,4,6,8 |
| **E** | 2,3,6,7,8 | 2,3,5,6,7,8 | 4 | 2,3,6,8,9 | 3,6,8,9 | 2,6,8,9 | 1 | 2,3,5,6,9 | 2,6,7,8,9 |
| **F** | 1,2,3,6,7,8 | 9 | 3,5,6,7,8 | 1,2,3,4,6,8 | 3,6,8 | 1,2,4,6,8 | 2,3,4,5,6,7,8 | 2,3,4,5,6,7,8 | 2,4,6,7,8 |
| **G** | 5 | 1,4,6,8 | 3,6,7,8 | 2,6,8,9 | 6,8,9 | 2,6,8,9 | 2,4,6,8 | 7 | 3 |
| **H** | 3,4,6,7,8,9 | 3,4,6,7,8 | 2 | 3,6,7,8,9 | 1 | 5,6,8,9 | 4,5,6,8 | 4,5,6 | 4,6,8 |
| **J** | 1,6,7,8 | 1,3,6,7,8 | 3,6,7,8 | 2,3,6,7,8 | 4 | 2,5,6,8 | 2,5,6,8 | 1,2,3,5,6 | 9 |

**Fig. 10.7** Example with First Pass List of Possibilities

- Again revise list, G3-9r and J6-5r&c [2 solved]:

  - B3 is set to '6' [1 solves]; and
  - B7-9 [1 solve].

- Again revise list, D8-9c and D9-4r [2 solved];
- Again revise list, D2-2r, E5-9c, F4-4r [3 solved];

  - B2-4 [1 solved];
  - B1-2 completes that row [1 solved];

- Again revise list, A8-2r, C8-4c, F8-3c, F9-2c, H1-4r&c [5 solved];
- Again revise list, A6-4r&c, D5-3c and D3-8c [3 solved];

  - F7-8, D7-6, E9-7 and F5-6 [4 solved].

- This then ripples through to solve:

  - C9-6, F1-7 and G5-8 [3 Solved];
  - C6-8 [1 Solved];
  - C4-9 and E6-2 [2 Solved];
  - C1-3 [1 Solved];
  - A3-7 and E4-8 [2 Solved];
  - G4-2 and C7-7 [2 Solved];
  - A2-8, E2-3, J1-8 and J3-3 [4 Solved];
  - A4-6, A7-3, E1-6, G6-6, H2-7 [5 Solved];
  - A1-9, J4-7 [2 Solved]; and
  - H4-3 and J2-6 (completing the puzzle) [2 Solved].

## 10.6  Puzzle Solvers

Although people are expected to solve puzzles without using computers, there are numerous tutorials tools available on-line. We have included some of the sites visited below:

- SudokuWiki.org[31],
- Sudoku Puzzles[32],
- 9 by 9 Sudoku Solver[33], and
- Sudoku Dragon[34].

A quick search on the net will also provide many downloadable options. Some provide step-by-step solving, others offer a selection of strategies, however at some

---

[31] http://www.sudokuwiki.org/sudoku.htm by Andrew Stuart.

[32] http://sudoku.com.au/.

[33] http://www.sudoku-solutions.com/sudokuSolver9by9.
phpfromAireTechnologies.

[34] http://www.sudokudragon.com/ through Silurian Software.

stage, a systematic trial and error process may be required (especially with AI Escargot)[35].

### 10.6.1 Heuristic Search

A heuristic method is one based on experience or training. This form of problem solving can be used to derive a solution that relates to the best possible answer. This is analogous to the term, 'an educated guess'. Generally a heuristic search is conducted using a 'trial and error' approach [289]. One could loosely contend people result to searching during a systematic trial and error process (although without assistance). A number of techniques have been developed in this field. Some include: Depth first, Breadth first, Hill climbing, Least cost, Annealed and A* searches. Unfortunately these techniques are not always deterministic and can stall under certain conditions.

### 10.6.2 Stochastic Force

Randomise the sequence of all blank values based on the remaining values in the series. Rotate these values until the error count reduces to zero. The search method used to determine the new sequence could rely on many techniques, some suggestions include *simulated annealing* or even *tabu search* [1, 133].

### 10.6.3 Brute Force Search

This is an elementary search based on the permutations of the existence of a solution within the problem domain [273]. Assuming there is one and only one valid outcome, a recursive search can exhaustively loop through all the values of a collection and exit this loop with the valid instance[36].

Using a well written *Brute Force* algorithm, a guaranteed solution will be achieved[37]. The algorithm is simple. The first digit/symbol is inserted into the puzzle and validated against the values already in the puzzle. If a conflict arises, the value is removed (using backtracking) and the next attempted. The process is recursive until the last piece of the puzzle is solved (In a 9 * 9 puzzle there are 81 values). Given the possible permutations presented, on a single core 3GHz processor, the initial software solution took between 30 and 45 minutes to solve. More optimised version of this code now exist.

Several computer programs exist that use a brute force algorithm that solve Sudoku puzzles. This method relies on progressively testing every square sequentially,

---

[35] Many of these use algorithms to invoke a number of strategies, however trial and error using backtracking will eventually be used.

[36] A flag can be used to indicate success or failure. An index can be returned to provide the location of a valid match. Where no solution is found the index value would be set to zero or null.

[37] Felgenhauer and Java calculated that there were 6.67 * 1021 possible solutions for a 3 * 3 Block or 9 * 9 numeral matrix in 2005.

row by row, until the puzzle is solved. The number of combination can compromise the speed and complexity of the solution. For example, most algorithms would start solving a puzzle using the values from '1' to '9' with a puzzle that represents a Non-deterministic Polynomial (NP) complete problem [415]. If the solution of that row was '987654321' the number of permutations compound the time required to solve a puzzle[38] [338]. Although time varies exponentially with complexity, however more advanced modifications could include algorithms using Dancing Links to solve puzzles in fractions of a second. With brute-force algorithms a solution is guaranteed (unlike stochastic, deductive or reasoning methods). The search methods may include simulated annealing and Tabu search [224]. Perez and Marwala successfully demonstrated the use of several stochastic search techniques: cultural algorithm, quantum annealing and the Hybrid method that combines genetic algorithm with simulated annealing [292].

### 10.6.4 Tabu Search

Glover improved on existing heuristic methods, by collecting *meta-heuristic data* to map out or exclude points within the solution space [133]. The search evolves in a cyclic manner that avoids retracing previous moves in order to reach a local minima. Given that new courses may not be chosen at random, efficiencies can be achieved. In other words, there is no point accepting or reversing a previously used (investigated) path that generated a poor solution. The concept of using *diversified* candidates to achieve an *intensified* process using measures associated with size, variability and adaptability of the *Tabu* memory [132]. This form of search is good when scheduling and routing, especially in an integer environment, such as a *travelling salesman* problem space.

## 10.7  Solving Sudoku Using Recursion and Backtracking

In view of the fact that all puzzles should be solved using logic, without search, you will be forced to use trial and error. A basic backtracking algorithm can be used to solve any valid Sudoku puzzle, although it can be slow and resource intensive. The premise involves testing the progressive substitution of values in a given order and backtracking upon failure (until complete or once all have been tried for a given vertex). There are many ways to optimize this form of algorithm. A simple graph tree can be used to navigate subsets with the total population. Using this approach searches can be used to sort the size, order or complexity of a series, prior to applying any algorithm. Figure 10.7 displays a puzzle that is relatively easy to solve by hand, however represents a 'Near Worst Case' where the solution begins with the series '9,8,7,6,5,4,3,2,1'.

---

[38] A puzzle with a low number of clues (17) on the top row with no clue values (all cells empty with a solution of '987654321'), could take between 30 and 45 minutes with a personal computer processor running at 3 GHz.

This technique involves a data structure (model), a GUI to view the results and a controller to implementing the algorithm. The same call can be used recursively to solve the problem and backtrack when it experiences an error in any of the row, column, blocks series tested. In this example the GUI will contain a matrix of buttons arranged in an $m * n$ grid of $m * n$ blocks, where $m = n$ and $m \geq 2$[39]. The problem can be preset in code, input manually by an operator or via a stream (scanned representation, photo, file, shared object or even a URL). Regardless of the input method, verification must be conducted to prevent runt-time errors.

AI theory can be used to provide solutions using many techniques. In the next chapter we will define our solution without search (as intended). Although one could create more efficient results by incorporating even the most basic AI techniques, like sort and search, our intention is only to use the puzzle concept to demonstrate agency theory.

---

[39] A matrix of one Block is possible, but impractical as are arrays greater than five blocks. Puzzle exist that represent $3 * 3$ (represented as numbers from 1-9), $4 * 4$ (represented as hexadecimal numbers 0-F) and $5 * 5$ matrix blocks (represented using alphabetic characters a-y or A-Y).

*"A physical symbol system has the necessary and sufficient means for intelligent action [276]."*

Allen Newell and Herbert A. Simon

# 11

# Problem Solving Workshop

In this chapter we work through a number of techniques raised in previous chapters. Each of the examples is accompanied with a brief explanation, although the syntax within some listings have been abbreviated to simplify the description. We introduced both static and dynamic methods associated with deriving results. A number of algorithms are then explained before discussing the MVC pattern associated with the Sudoku solver application.

## 11.1 Sudoku Puzzle

We have encountered many websites promoting this strategy based puzzle, such as *Sudoku Dragon, EzSudoku* and *Sudoku Essentials*. Most provide detailed descriptions on strategies to solve puzzles using a variety of techniques.

J.W. Tweedale & L.C. Jain: Embedded Automation in Human-Agent Environment, ALO 10, pp. 163–176.
springerlink.com                                        © Springer-Verlag Berlin Heidelberg 2011

## 11.2   Problem Solving

There are many ways to solve problems. Over the past century, humans have creates progressively more complex toys and puzzles. Examples include simple peg board, jigsaws, card games, board games, computer games and now even electromechanical entertainment devices. Technology has played an important role in this evolution. Both AI and CI have featured prominently as problem solving tools. We loosely categories these tools based on the technique used to solve or optimise each problem. These generally include problem solving using:

| | |
|---|---|
| Search: | Brute force thru A*; |
| First Order Logic: | Agents, Knowledge Inference, Planning and Reasoning; |
| Uncertainty: | Combinations, Permutations, Probabilities and Decision Making; |
| Learning: | Case Based, Guided (Knowledge Based), Probabilistic and Reinforcement; and |
| Perception: | Natural Language, Interoperability, Communications, Cognitive Perception and Robotics[1] [320]. |

There are many courses in universities that teach problem solving with computers. They begin by identifying skills and personality traits of humans before moving into heuristic methods. Most people initially approached problems using a *Trial and Error* approach[2]. To provide a practical example, this text will follow the approach. A number of strategies that could be implemented are discussed below. The methodology that aligns to this approach relies on recursion for the guess and backtracking to correct any errors or resolve conflicts. Generally all problems are solved within a constrained environment based on a preset number of conditions, such as position, state or orientation. The biggest issues generally encountered the humans is the interface required to initiate or maintain those conditions. We have chosen to use a tradition window based presentation to dynamically display the puzzle as algorithms permeate towards the solution.

The human interface is more accepted when implemented as a point and click (or touch) display. In this case a Soduko matrix can be displayed using a GUI show as a matrix or grid of buttons. The buttons will display the current state, can react to human input and can be used to trigger the logic required to process the interaction. This code can also verify and validate the type, range and accuracy of the input (this may eliminate conflicts and reduce the complexity of the code by confining the inputs to known values).

---

[1] This edition is being completely revised for release as the third edition.

[2] Nobody reads the manual anymore, they simply push buttons until the desired goal is achieved.

## 11.3 Simple Grid Display

Listing 11.1 show how to create a basic frame with an *'m' * 'n'* matrix of buttons. Examples of this display are shown in Figure 11.1. As shown, the work can be done traditionally using the *BackTrakingSolver* class. Please note the syntax may be abbreviated in order to assist in simplifying some of the example listings.

```java
public class test extends JFrame {

    public static void main(String[] args) {
        // BackTrakingSolver bts = new
                BackTrakingSolver();

        // Create and set up the window.
        JFrame frame = new JFrame("Symdoku Puzzle ");
        frame.setDefaultCloseOperation(JFrame.EXIT_ON_CLOSE);
        frame.setSize(new Dimension(300,300));

        // Create and set up the new content pane.
        Container pane = frame.getContentPane();
        pane.setLayout(new GridLayout(9,9));

                // bts.createModel() ;
        // bts.createView(pane) ;

        // Display the window.
        frame.pack();
        frame.setVisible(true);
    }
}
```

**Listing 11.1** Create Puzzle model

We don't intend on providing a detailed explanation of our code, however we will describe the general flow. Our main window uses code that is common to many Java GUI containers. This window is used to display (view) and control the overall application. It can be customized and decorated with little effort. Once instantiated, it is displayed (set *visible*). Once visible, a nine by nine grid of buttons will be presented as a matrix of three by three grids/blocks. At this stage all the buttons will have the user input disabled, although they will be updated programatically. A description of this model follows.

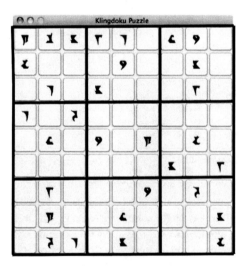

(a) Numeric Puzzle

(b) Klingon Puzzle

**Fig. 11.1** Example of the Symdoku GUI

## 11.4   Algorithms

As with all problems, a number of algorithms can be used. Initially most researchers use brute force algorithms to provide a workable solution. Once the concept is proven, follow-on research is generally conducted on alternative methods. The base solution is used as a benchmark upon which improvement are measured. It is assumed that any optimization is represented as an increase over the base readings. We are not claiming any improvements or optimization. We do however believe that our technique can be used to provide more efficient and streamlined heterogeneous applications that are capable of working in a distributed environment.

### 11.4.1   Create the Model

Listing 11.2 displays the basic outline of how we create our model. For simplicity, you can define the values manually by initialling Where $p$ is the number of cells across the matrix. As discussed in chapter10 we have decided to create a symmetrical puzzle with an equal '$m$' * '$n$' matrix, given '$m$' = '$n$'.

```
1     protected void createModel ()
2     {
3      model = new int[p][p] ;
4
5      // Clear all cells
6      for( int row = 0; row < p; row++ )
7         for( int col = 0; col < p; col++ )
8            model[row][col] = 0;
9
10     // Create initial values here using the
11     // row/col convention as a (0-8) array.
12     // model[0][0] = 1;
13
14     // additional values . . .
15
16     // model[8][8] = 9;
17
18    }
```

**Listing 11.2** Create Puzzle model

### 11.4.2   Update Algorithm

Our controller requires code that updates the puzzle when values change. We take a simple approach and update every value within a loop (see Listing11.3). This can be achieved using the method that created the data model. The only variation is the use of a null (blank) in lieu of displaying '0' for unknown or unsolved values (starting line 7). We could have used any character (such as a '?'), but the puzzle looks cleaner this way.

```
1    /** Updates Table view from the model */
2    protected void updateView()
3    {
4        for( int row = 0; row < p; row++ )
5            for( int col = 0; col < p; col++ )
6                if ( model[row][col] != 0 )
7                    view[row][col].setLabel( String.valueOf(
                            model[row][col]) );
8                else
9                    view[row][col].setLabel( "" );
10   }
```

**Listing 11.3** Updates the Values within the Table View

### 11.4.3  Value Checking Algorithm

Listing 11.4 displays the algorithm for the checks performed within selected Rows. The same approach is used to check the values for Columns and Boxes. You simple alter the variables to navigate the Column or Box. The passed value is compared with the current value in line 5.

```
1    /** Checks if num is acceptable for the given row */
2    protected boolean checkRow( int row, int num )
3    {
4        for( int col = 0; col < p; col++ )
5            if ( model[row][col] == num )
6                return false;
7
8        return true;
9    }
```

**Listing 11.4** Check for Acceptable Value

### 11.4.4  Solve Algorithm

Lisiting 11.5 shows the basic steps required to iterate through the puzzle until it is solved. Please note, you must provide a puzzle with only one valid solution. This is tested in line 19. If more than one solution exists, the algorithm returns the first solution encountered.

```
1    /** Recurse to find a valid number in a single cell */
2    public void solve( int row, int col ) throws Exception {
3        //add delay to allow users see the change
4        System.out.println("Start Solve Process");
5        pause.delay(1000); //show original puzzle
6
7        // Stop the process if the puzzle is solved
8        if( row > (p - 1) ) {
9            System.out.println("Solution found");
10           throw new Exception("Finished"); //exit stack
11           }
12
13       // If the cell is not empty, continue to next cell
14       if( model[row][col] != 0 ) {
15           next( row, col ); }
16       else {
17           // Find a valid number for the empty cell
18           for( int num = 1; num < (p + 1); num++ ) {
19               if( checkRow(row,num) && checkCol(col,num) &&
                     checkBox(row,col,num) ) {
20                   model[row][col] = num;
21                   updateView();
22                   next( row, col );
23               }
24           }
25
26           // No valid number found so clean up & return
27           model[row][col] = 0 ;
28           updateView() ;
29       }
30   }
```

**Listing 11.5** Solve Algorithm

## 11.5  User Interface

The class outline of the Sudoku agent puzzle application is depicted in Figure 11.2. When running as a concurrent thread, a MVC enables that control to listen to changing stimuli (as a parallel process is solving the problem).

## 11.6  Solution to Puzzle

As discussed in Chapter 10, the example provided in Figure 10.7 is difficult for computers. Not as difficult as *AI-Escargot* (see Figure 10.6) but as suggested, is *Near Worst Case*. There is a solution as shown in Figure 11.3. Surprisingly, it is relatively straight forward for humans to solve and with a minor optimization of our agent, we could easily make it simplistic for the computer to solve. One approach is to rotate

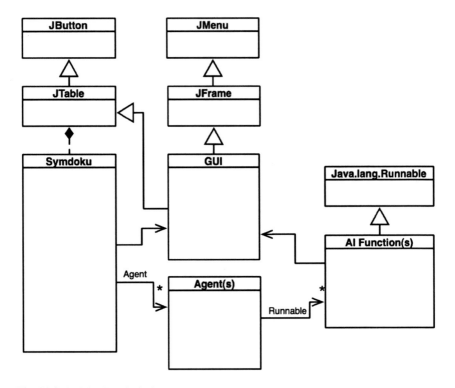

**Fig. 11.2** Sudoku Puzzle Software

the puzzle by 90 degrees (transform it programatically). The figure shows the nines shown in A1 and J9. These would translate to cells A9 and J1 making the puzzle solvable by a computer in seconds. A true agent system could be wrapped with optimization techniques. A supervisor could launch multiple agents and use a voting system to determine the best outcome. The status of each agent could also be preempted to manage the overall system performance and efficiency.

## 11.7  Java Transforms

The instrument package can be used by the java Agents library. Given the open-source support of Byte Code Engineering Library (BCEL) [73] and refers to the asm keyword (ASM)[3] [39], developers can instrument their Agents accordingly, through the use of the *premain* bootstrap class. Agents must be packaged in *jar* files and attached to an instantiated application. The syntax for the *manifest* files is strict and can be a prime source of errors, however libraries have been created to automate

---

[3] The goal of the ASM library is to generate, transform and analyze compiled Java classes, represented as byte arrays.

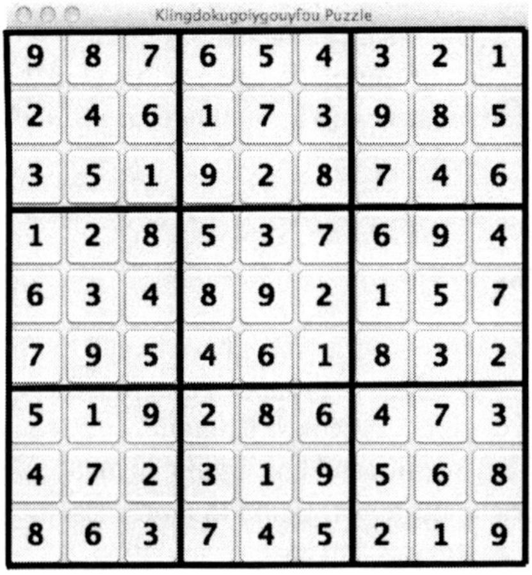

**Fig. 11.3** Puzzle Solution

this task. This needs to be part of any *agent factory*. Once created, agents can be attached to a running application in an existing JVM from the command line using the *-javaagent: jarpath[options]*, it can also be launched within a JVM using the instrumentation class (*java.lang.instrument*). Either way, the agent is loaded by the class loader, which will currently ignore any uncaught exceptions.

An agent can be written to run as a stand-alone body of code. It can also be attached to an existing application in memory at run-time (such as a user interface). You can also transform or instrument your class as it is loaded (run-time). This capability has existed since Java 5 and commonly occurs when using AOP, EJB or any other light-weight containers [286].

Using the *Java Transformer Class* a programmer can either add a *fascard* to an existing class at *run-time* or register a file transformer for use when called. It is presently an extremely manual process and requires very strict syntax, packaging and access to the parent in memory [333]. The value of this class is the improved ability for testing multi-threaded applications. This tool and these concepts can be embedded into a class to automate the build process dynamically, example code exists within Apache Software Foundation (ASF) and Inria [416] development platforms.

## 11.8  Agent Programming Language

The concept of an agent programming language is similar to the approach used in an agent factory. The syntax and semantics would resemble that of a traditional language. A library of agent capabilities can be initiated using environmental stimuli and progressive results passed dynamically as the solution evolves. Design, coordination and control become significant challenges. A crude architecture similar to early microprocessors (Op codes/capabilities) could be created with alternative (extension) calls provided using wrapper classes. This concepts becomes extremely complex unless agent classes can be dynamically adapted. As discussed in section 9.10, we cite the use of **premain()** in lieu of **main()** to achieve this goal. Using this approach, an agents can be dynamically attached to a running process, such as the interface used to view the puzzle.

### 11.8.1    Attach Algorithm

In order to use agents in Java, programmers have to call the *transformer* class to attach it to a running process. Given the scope of that process, methods can be created to traverse the structure; either as Rows, Columns and Blocks. The problem scenario will be loaded using a command line argument that provides the name of a file containing the appropriate attributes. Using this process, the method can be easily extended to include other streams and enable validation. As suggested the *Attach* algorithm can be used to launch the agent after acquiring a reference to the puzzle GUI. Listing 11.6 shows how we used the *sun.tools.attach* library in conjunction with the current threads PID to dynamically attach a valid agent.jar file to it while in memory at run-time.

```
 1  import com.sun.tools.attach.*;
 2
 3  public class Attach {
 4      public static void main(String[] args) throws
            Exception {
 5
 6          if (args.length != 2) {
 7              System.out.println("usage: java Attach ");
 8              System.exit(1);
 9          }
10
11          // This program accepts two parameters:
12          //    1. The pid of the JVM on which to load an
                   agent
13          //    2. The full path of the agent jar file to
                   load
14
15          // JVM is identified by process id (pid).
16          VirtualMachine vm = VirtualMachine.attach(args
                [0]);
17
18          // load a specified agent onto the JVM
19          vm.loadAgent(args[1], null);
20      }
21  }
```

**Listing 11.6** Using the Sun Attach Tool

## 11.8.2 Basic Agent Structure

Listing 11.7 displays the basic structure required to create an instrumented agent class. The use of the *agentmain* keyword serves as the main function called by the instrument class after attaching the packaged class during run-time. We could make the example as simple or complex as required.

We have chosen to use a basic *Hello World* example. A print statement echos our message, although we do provide additional code to enable fault finding of the syntax. The most common mistakes made are excess carriage returns and/or unnecessary white space. In this case, as simple message is displayed. In this case we echo the arguments passed in on the command line.

```
1  import java.lang.instrument.*;
2
3  public class HelloWorldAgent {
4      /* NOTE: agentmain is the special method that
5         will be called when this agent is loaded */
6      public static void agentmain(String agentArgs,
           Instrumentation inst) {
7          System.out.println("Hello agent!");
8          if (agentArgs != null) {
9              System.out.println("my args are: " + agentArgs
                );
10         }
11     }
12 }
```

**Listing 11.7** Basic Agent Class

To demonstrate this concept, build and run the test.java file and use the attach.java to attach the agent.jar class when the test file is running. Listing 11.8 show displays a simple loop that introduces an active thread within the JVM. The active PID can be obtained by running JPS or Top. The jar file can be built using Listing 11.9 and run using the command in Listing 11.10. The output produced is shown in Listing 11.11. Each time you run the attach script, the message is repeated. To end the *Test* class you need to either break or kill the thread.

```
1  public class Test {
2      public static void main(String[] args) throws
           Exception {
3          System.out.println("Main...");
4          while(true);
5      }
6  }
```

**Listing 11.8** Test Class

```
1 jar cvfm helloAgent.jar manifest.mf HelloWorldAgent.class
```

**Listing 11.9** Build Script

```
1 java -cp /classpath/tools.jar:. Attach 2108 /path/
      helloAgent.jar
```

**Listing 11.10** Attach Script

```
1 Home[path]: java Test
2 Main ...
3 Hello agent!
```

**Listing 11.11** Output Displayed

Once you achieve a successful build, you can start including more complex effects, like changing the value of variables or enhancing the default class behaviour (such as logging and mobility discussed previously). This research is ongoing with many open source projects being created to champion the spread on enterprise level capabilities. The examples in this chapter provide the working knowledge required to get started. The next chapter summarises our research and outlines our future aims.

## 11.9  Suggested Exercises

For the beginner it would be easier to start with fewer variables. Take a problem like tic-tac-toe. This is a common turn-based problem played by two opponents on a 3*3 grid. They progressively place a nought '0' (zero) or cross 'X' in random cells in an attempt to be the first to achieve three marks in a row, column or diagonal. The game is over when all nine cells are filled. The game is drawn if neither player achieves three successive marks. A common strategy used is to block the opponent to acquire a draw, rather than a loss.

1. Calculate the combinatorial probability for placing '0' on the board. There are 512 combinations possible for the entire board 9! And only 5! for the first players symbol and 4! for the second players symbol or C(9,5).

2. Draw all nine ply's of an two-player extensive-form game tree for either player making the first move.

3. Design a look-ahead strategy to ensure you never lose (assume you are the second player using crosses 'X'). The first player places the '0' randomly in any cell as the first move.

To expand this concept, consider the n-queens problem. Here you are asks to place 'n' queens in a 'n*n' chess board, such that no pair of queens can attack each other based on the rules of standard chess. This represents Constriaint Programming (CSP) (Constraint Satisfaction Problem) and is considered a non-trivial problem. There are eight rows and eight columns on a standard chess board or 64 possible locations for each of the eight queens. However there are only 92 possible solutions (64*63*62*61*60*59*58*57/8!). For a small array (N<9), brute force is computationally feasible, but as the number of alternatives grow, heuristic techniques are required. Remember by introducing a number of constraints you con minimise the number of guesses. Consider implementing a solution using:

- No two queens can be placed in the same row,
- No two queens can be placed in the same column, and
- No two queens can be placed in the same diagonal.

4. Create a simple algorithm to solve a problem with only four queens.

5. Develop a heuristic algorithm to enable you to pragmatically solve a problem with 15 queens.

6. Try developing a simple backtracking algorithm using the above constraints.

7. If you are adventurous, look at developing an algorithm that can solve this problem using parallel processes. This solution would be easily embedded into an FPGA.

## Further Reading

G. F. Luger. Artificial Intelligence. Addison-Wesley, Reading, MA, 5th edition, 2005 [229].
O. Dahl, E. W. Dijkstra, and C. A. R. Hoare. Structured Programming. Academic Press, London, 1972 [71, pp. 72-82 (Dijkstra's 8 Queens solution)].
J. J. Watkins. Across the Board: The Mathematics of Chess Problems. Princeton University Press, Princeton, N.J., 2004 [392].

*"What thinks is a brain in a human being who is part of a system
that includes an environment [18]."*

Gregory Bateson

# 12

# Future and Direction

The research in this book examines the origin of MAS and demonstrates how they can
provide autonomous *capabilities*. One or more facilitator agent can be used to coordinate the execution of a decomposed task by a team of agents employed to process
dynamically acquired actions. This acquisition can provide flexible re-use at run-
time, without the need to re-instantiate agents or their capabilities. Agent teaming
techniques have been used to enhance the behaviour and flexibility of agent communication using polymorphism. Through collaborative efforts, rudimentary MOE or
MOP measures have been devised to resolve trust issues. However this research is
ongoing. Any questions about efficiency and scale have been delayed until the model
has matured or other researchers adopt the challenge.

A number of Agent Factory Demonstrator (AFD) models have been developed at the University of South Australia that involve leading edge concepts
which are intended to exploit the dynamic creation of agent capabilities with fast
and efficient context switching. The simulator concentrates on threads (as apposed to processes), it is programmed in Java, with a GUI that displays the state
and results of all agents in the system (on the originating system or using distributed techniques on one or more connected machines). As discussed a survey
explains that existing communication languages used with agency software are limited in functionality. That is why SOAP has been chosen for use in our future
demonstrator designs. Given time and interest from other researchers, the challenge of building a multi-lingual universal translator capability for communication and tasking is planned. Existing research has already established architectures

J.W. Tweedale & L.C. Jain: Embedded Automation in Human-Agent Environment, ALO 10, pp. 177–204.
springerlink.com                                                 © Springer-Verlag Berlin Heidelberg 2011

involving JACK, JADE and CIAgent cannot achieve these goals efficiently in isolation. Solutions to those lessons are being adapted for use in the AFD. Further work is required to mature this design, however the results so far are very encouraging. Technology, software, interoperability and dynamic functionality remain the major challenges for researchers attempting to improve applications in this paradigm.

## 12.1   Innovations in Modern AI

In Chapter 2 we discussed the issues surrounding the evolution and science surrounding the birth of AI and its evolution as CI. There has been a transition of ideas and research to explain the definition of what AI really means. Researchers continually struggle to prove that the *Turing Test* demonstrates CI. Minsky indicated that intelligence represented *physical syntactic symbolic manipulation*, but Searles *Chinese room* demonstrates *the whole cannot be more than the sum of the parts*, and *complexity theory* extends the definition of what constitutes the *whole sum*. We understand that AI is NOT what is being portrayed in current cinematic phantasies and provide an historical perspective leading up to MAS and which directions it may lead in the future.

## 12.2   Direction of Autonomy

We began by introducing existing research into Knowledge-Based engineering and Information Systems to provide automation and innovation. In order to outline this topic we clarified the definition of the term intelligence and the tools required to exploit human machine interaction. After discussing the assumptions and describing how knowledge can be represented we provided discussion on how human factors and decision support systems can be used to make decisions within computer systems. Decisions are the key to automations, but traditionally machines are controlled by human operators and unless the interfaces enable seamless operation to maximise efficiency, they remain reliant on the next input. We promote human-agent teaming by discussing the concepts of a human centric approach and the requirement for man-machine coordination before describing next generation systems.

## 12.3   Agent System Frameworks

The science of AI gathered significant momentum following the introduction of computers to break codes during the second world war. Samuels used the opportunity of using computer to computationally project winning moves in the *checkers* game while working at International Business Machines (IBM). We briefly revised how technology stimulates the growth of AI and the tools used to solve real-world problems. We also systematically explore various types of architectures used to promote modern AI. For instance, OOPL has evolved sufficiently to enable programmers to rapidly produce platform independent code with significant interoperability and re-use. A number of standards, libraries and capabilities with Plug and Play (PnP) frame-works now exist that enable a diverse range of programmers to code applications with millions of instructions in a relatively short period. As discussed in Chapters 3 and 4, design tools, patterns and component technology assist in accelerating this process further. The largest impediment with coding software is creating efficient applications. Even multi-core microprocessors rely on processes that are managed sequentially, resulting in significant wait time when processes are blocked to service those with higher priorities.

Two vendors (IBM and Apple) have acknowledged that they are developing commercial tools which assist programmers in writing parallel process applications. Apple are extending these tools by offering dynamically access to any idle CPU time available on graphic chips, physics engines and other accessible hardware. The next major issue to solve is that of interoperability and the dynamic delivery of functionality. OOPL and OSs have been developed with a variety of features, however many extend or wrap existing libraries with enhancements. As with technology, vendors and programmers are resorting to mainframe techniques to create enhancements or increased efficiency, making most process unnecessarily convoluted. Todays applications need to be dynamically scalable, across distributed networks using secure access via responsive, flexible communication protocols that are reliable, robust and fault tolerant. To achieve this the next challenge is to redesign the OS enabling more efficient thread development and distribution. This is not the prime topic of this text, however like Interoperability, the general theory needs to be taken into context to make an informed judgement about the true scope of the future challenges facing the CI community.

## 12.4 The Role of Autonomy in Agents

Prior to discussing autonomy it is important to review the terminology, syntax and semantics surrounding this topic. As discussed in Chapters 5, we cover Topology, Ontology, Taxonomy and OSI before launching into the requirements for communication protocols. It is important to reflect on the relationship between standards, protocols and the applications attempting to implement both. We attempt to align several of these concepts in Figure 5.1. we discuss the basic concepts about agent communication languages, including ACL, KIF, KQML, FIPA and SOAP. Again we briefly examine a variety of DARPA Agent Markup Language (DAML) and the use of WSDL/XML. These tools are becoming vital in enhancing information exchange technologies used in autonomous systems.

## 12.5 Enhancing Autonomy

The concept of MAS autonomy was defined by Wooldridge and Jennings [408]. Without the ability to make decisions, Agents become processes, functions or dedicated sequences of code that can be written and monitored in various guises in generic applications. Given autonomous characteristics, agents can gather data, make decisions and generate a suitable COA based on the environment and its tasking. Learning, context repositories and the ability to dynamically adapt the functionality of an entity are all ideal enhancement that are not commonly available to teams, MAS or even systems of systems approaches to software development. The AI community does have access to mobile agents for Web based applications, however very few are efficient, communicate effectively or with the desired response expected at a system level. To achieve the scale demanded by many corporate customers these capabilities must be provided synchronously across distributed systems.

As discussed in Chapter 6, when discussing agents, the definition of autonomy is still evolving. Similarly the term interoperability has a maturing definition. The integration of two agents in a stand-alone application can be controlled and coded/edited at design time and tested at run-time to ensure that it operates seamlessly. This approach works until the same code is re-used, becomes subject to variations in context or an expanding problem space. Standards have been developed for a number of languages and the programmer can compensate for divergent ontologies using self healed techniques like XML, however if the structure, interface and functionality don't align completely problems will occur, compromising the data and eventually causing run-time errors. A designer/programmer cannot plan or test every possible condition that could occur in the problem space. The absence of even the most element of data could be detrimental, especially with safety/mission critical applications. Entities, based on agent technologies, need to be adaptive and possess the ability to dynamically react to the changing environment/context. The architecture needs to be capable of providing functionality during run-time on call to cater for unknown situations and be intelligent enough to be able to hand-off processes/plans it is unable to complete. Learning, planing and re-planning are important aspects of providing autonomous teams that have access to reconfigurable components and functionality. This hierarchical structure needs to be able to communicate with peers, siblings and supervisors, making agents the ideal building blocks when creating dynamic systems. Technology plays a key role in how these systems behave, communicate and interoperate. Modern programmers are required to understand the concept of multi-thread applications, composed from an aggregated source of components or services. We discuss ATLAS and TNC architectures before describing a generic application incorporating a UAV scenario.

## 12.6   Adaptation and Synchronous Functionality

Technology has evolved over the past century. Fixed bulky and cumbersome steam engines have been replaced with portable, miniaturised kinetic coils, fuel cells and batteries. Mechanical actuators now use precision digital encoders that are fused in modern high-speed, distributed computing systems. Data is gathered and manipulated to ascertain knowledge from its origin. Examples include fuzzy logic systems used to position train carriages to ensure doors are opened at specific locations, images are being scrutinized against one or more key frames to decipher patterns and images. Bayesian and Neural networks are being used to classify data which can be submitted to a number of follow-on CI techniques.

Most of the techniques used in AI have software solutions and many successfully committed to silicon using ASIC, FPGA and dedicated logic circuits. The suggestion using an FPGA solution could practically behave dynamically by re-configuring the firmware of a chip connected to operate as a co-processor (on-board), as part of a daughter board, mezzanine card or via a distributed interconnected device or system. Over the past decade industry has used increased parallelism in order to maintain Moore's law. Previously, speed increases were used to create actual gains in processor performance. Today virtual gains are reported as a result of providing more microprocessor cores. As discussed, a CPU based on the von Neuman architecture would need to run at over 25 GB/s in order to create a viable synapse. The resulting issue common to all of these factors is that existing microprocessors do NOT support efficient solutions for many of the existing AI techniques and until this bottleneck is corrected, the concept of artificial entities will continue to languish. The minimum enhancement suggested include indexed banks of register rich memory belonging to specific threads which can be switched or transferred among another bank of CPU cores. The inclusion of dedicated high density memory arrays interconnected by programmable logic bridges to alternate banks of high density memory that can be migrated between cores is also desirable. Again other banks of re-configurable logic that could be used for Formal, Bayesian and Fuzzy logic processing. The gamete of hardware solutions is broad and like autonomy is not the prime topic of this text.

Chapter 7 discusses the rise of physical and virtual architectures, including the concept of abstracting software into silicon. We then integrate the issues raised throughout this text. As discussed there are many issues surrounding AI, agents, teaming and autonomy (in both software or hardware). Programmers encounter this complexity when attempting to decompose problem statements. We acknowledge that since the industrial revolution the only certainty in our lives is change. Research evolved and a myriad of domains expanded, especially during times of conflict, war and periods of strained foreign policy. Most scientists choose to remain ignorant of these events or actively assist in creating solutions to one or more of the challenges presented. The British stimulated research in computer science out of the need to solve cryptographic messages. This stimulated the spread of AIP techniques using automated machines. CI developed during this environment and many sub-domains resulted. Under the guise of network centric warfare, the goal is to create a system-of-systems approach that dynamically adapts to the situation via a series of interoperable framework of synchronously adaptable elements within the distributed network of assets being employed at any single point in time. The TNC model discussed illustrate one solution to providing improved communications based on a trust mechanism that promotes self discovery using a predefined series of seamless interoperable protocols. Other demonstrations discussed illustrate the ability to incorporate design patterns, thread pools and polymorphism to achieve dynamic functionality in response to the environment of changes in context. More research on these topics is required before being applied to a real-world problem. This will be achieved using a blackboard demonstration environment that provides modularized components which will use common interfaces to deliver a variety functions or capabilities. This distributed simulation environment will enable students, whether part of KES or participating in collaborative research, to conduct controlled experimentation with the ability to make comparative measurements to ascertain efficiency, throughput and latency of these concepts in a controlled problem space. The tools provided will also enable future researchers to fast track alternative algorithms, assumptions or scenarios. We further offer discussion on dynamic architectures and even project the concept of micro-simulated capabilities that integrate forward synthesis to stimulate model (agent) based computation techniques.

## 12.7   Agent Oriented Programming

There are many architectures and technologies or domains of influences that have been used to enhance agent systems. When implemented as a system, agents are capable of achieving highly sophisticated goals autonomously. Each capability can be written to interoperate seamlessly based on standard interfaces and protocols. The focus on autonomy and migration has continued to the point where packages of work (sub-tasks) are being centrally coordinated. Parent agents are now capable of scripting capabilities based on procedural calls similar to *op codes* experienced by programmers using assembly language. In Chapter 8 we investigate the evolution of agents architectures, their level of control, construction and mobility. We discuss the evolution of agents prior to exploring frameworks like *Aglets*, *CAST*, *JACK*, *JADE* and *SOAR*. Discussion continues to explore communications and how data is passed or exchanged. There has been a paradigm shift to embrace more hybrid systems, as demonstrated by NASA. Agents now migrate to the information source to process queries or tasks. Once satisfied they can proceed to subsequent sources for further processing or simply return with the requested solution. The industry has embraced the use of *models*, *patterns*, *proxies*, *generics* and *multi-threading* to satisfy interoperability. Many of these techniques require knowledge and expertise to cope with the complexities presented in achieving flexible systems. Agent factories can be used to abstract many of the issues encountered.

## 12.8   Creating an Agent Factory

The concept of an Agent Factory was discussed in Chapter 9. Given that we are writing this book to support students and researchers, we choose to focus on Java as a development environment. As such we believe a review of its origins and recent enhancements was appropriate. There is significant effort to improve its image and align the core capabilities closer to that of C++ through the JSR program. These enhancements do effect what and how improvements can be achieved. Most of this discussion surrounds ways of interacting with the JVM, either locally or throughout a distributed heterogenous environment. This involved discussion about gaining *scope* using *proxies* to provide more dynamic agency behaviour. Upon establishing the tools, we discussed the interfaces and the ability to package or migrate agents.

## 12.9 Sudoku Case Study

Over the past 50 years, AI research has been demonstrated using puzzles and games. Chapter 10 conducts a case study on the problem of solving a sudoku puzzle. We introduced the terminology, representation and mathematical analysis. Discussion evolved to include the rules, theory and strategies, from novice to advanced techniques. To help the reader in understanding some of these strategies, we included a walk through that progressively solved each vacant cell. We don't intend on using AI to optimise our solution using search, but have provided a brief introduction of several options. We choose to solve this problem using recursive backtracking algorithm.

## 12.10 Problem Solving Workshop

We choose to use the Sudoku puzzle to demonstrate our theories because it is very popular and can be solved using a variety of AI techniques. We insisted on using a technique that maintained the original aim of the game to simply apply strategy and NOT search. Chapter 11 provides a workshop acts as a constructive guide to the techniques used to solve the puzzle. We discussed how the data should be represented and the use of an MVC approach to the HMI application. The display provides a simple grid of rows and columns, collectively arranged into blocks representing a '9*9' array of cells. The solving algorithm simple looped through each of the series constraints to determine the validity of available values selected. The checking algorithm rejected invalid or conflicting values with reference to rows, columns and blocks in accordance with the rules. This problem can be solved without using agents or any dynamic functionality. We provide an introduction to the Java transformation class and one method to attach agents to an active process in memory. The attach algorithm relies on being able to obtain a PID to the active process. As shown a minor modification enable programmers, or an agent factory, to package and attach any capability during run-time. Discussion in this book includes the ability to give the resulting agent mobility.

## 12.11 Looking Ahead

Many researchers would agree that many problems in this domain have been solved, however most solutions remain piece-meal or focus on a specific problem. Most solution apply to the WWW framework, based on concepts derived using client/server demonstrations. Although commercial applications in the form of shop fronts exist, many related topics require significant effort before being accepted by mainstream industry. Corporations are now demanding distributed, efficient applications that can reliably scale on demand in a secure environment with all the convenience presented by the WWW ideology. Ongoing research like ATLAS, TNC, the KES Black-box

Model and future concept demonstrators, all strive to deliver that and more. As discussed in the body of this text, more changes in AI research are required. The architecture of past and present hardware fails to align with the requirements needed to produce efficient or seamless CI applications. Although software has evolved, it is becoming layered, reducing flexibility, efficiency and the ability to dynamical adapt to the changing environment. Security, distribution and communications also remain impediments to future development. However trust, learning and present OSs still require significant effort to achieve autonomous interoperability using dynamic agent teaming functionality.

# References

[1] Aarts, E.H.L., van Laarhoven, P.J.M.: Local search in coding theory. Discrete Mathematics 106–107, 11–18 (1992)

[2] Alexander, C., Ishikawa, S., Silverstien, M.: A Pattern Language: Towns, Buildings, Construction. Oxford University Press, Carey (1977)

[3] Alterman, R., Bookman, L.A.: Reasoning about a semantic memory encoding of the connectivity of events. Cognitive Science 16(2), 205–232 (1992)

[4] Aman, U.: On Code Generation in a Pascal Compiler. Software—Practice and Experience 7(3), 391–423 (1977)

[5] Andresen, S.L.: John McCarthy: Father of AI. IEEE Intelligent Systems 17, 84–85 (2002)

[6] AOS. JACK Intelligent Agents: Teams Manual 3.0. Agent Oriented Systems, Melbourne (2004)

[7] AOS. JACK Intelligent Agents Teams Manual 4.1. Agent Oriented Services, Melbourne (2004)

[8] Appleton, B.: Patterns and Software: Essential Concepts and Terminology (2000), http://www.bradapp.net/ (Last Accessed March 26, 2007)

[9] Asadi, N.B., Meng, T.H., Wong, W.H.: Reconfigurable computing for learning Bayesian networks. In: FPGA 2008: Proceedings of the 16th International ACM/SIGDA Symposium on Field Programmable Gate Arrays, pp. 203–211. ACM, New York (2008)

[10] Austin, J.L.: How to Do Things with Words. University Press, Oxford (1962)

[11] Avancini, H.: A Java Framework for Multi-Agent Systems. SADIO Electronic Journal of Infomatics and Operations Research 3(1), 1–12 (2000)

[12] Babbage, C.: The Economy of Machinery and Manufactures, 2nd edn. Project Gutenberg Literary Archive Foundation, USA (1832)

[13] Balbo, F., Pinson, S.: A transportation decision support system in agent-based environment. Intelligent Decision Technologies 1(3), 97–115 (2007)

[14] Banerjee, D., Tweedale, J.: Reactive (Re) Planning Agents in a Dynamic Environment. In: Shi, Z., Shimohara, K., Feng, D. (eds.) IFIP International Federation for Information Processing, Intelligent Information Processing III, Adelaide, Australia. Computer Science, vol. 228, pp. 33–42. Springer, Boston (2007)

[15] Banks, S., Lizza, C.: Pilot's Associate: a cooperative knowledge-based system application. In: Intelligent systems and their applications, vol. 6(3), pp. 18–29. IEEE Expert, NY (1991)

[16] Barr, A., Feigenbaum, E.: The Handbook of Artifcial Intelligence, vol. 1. Morgan Kaufmann, Los Altos (1982)

[17] Bates, J.: The role of emotions in agents. Communications of the ACM 37, 122–125 (1994)

[18] Bateson, G.: Steps to an Ecology of Mind. University of Chicago Press, Chicago (1972)
[19] Belhajjame, K., Embury, S., Paton, N., Stevens, R., Goble, C.: Automatic Annotation of Web Services Based on Workflow Definitions. ACM Transactions on the Web 2(2), 1–34 (2008)
[20] Bent, R., Hentenryck, P.V.: Waiting and Relocation Strategies in Online Stochastic Vehicle Routing. In: Veloso, M.M. (ed.) Proceedings of the 20th International Joint Conference on Artificial Intelligence, IJCAI 2007, Hyderabad, India, pp. 1816–1821 (2007)
[21] Benz, B., Durant, J., Durant, J.: XML Programming Bible. Wiley, New York (2003)
[22] Bergenti, F., Poggi, A.: A Develoopment Toolkit to realize Autonomous and Inter-Operable Agents. In: Fifth International Conference of Autonomous Agents, Montreal, pp. 632–639. ACM, New York (2001)
[23] Berthier, D.: The Hidden Logic of Sudoku. Lulu, Paris, France (2007)
[24] Bichindaritz, I., Marling, C.: Case-based reasoning in the health sciences: What's next? Artificial Intelligence in Medicine 36, 127–135 (2006)
[25] Bigus, J.P., Bigus, J.: Constructing Intelligent Agents Using Java. Professional Developer's Guide Series. John Wiley & Sons, Inc., New York (2001)
[26] Billings, C.E.: Human-Centered Aircraft Automation: A concept and Guidelines. Technical Memorandum 103885. NASAAmes Research Center, Moffett Field (1991)
[27] Billings, C.E.: Aviation Automation: The Search for a Human-Centred Approach. Lawrence Erlbaum Associates, Inc., USA (1997)
[28] Blackmore, S.J.: The Meme Machine. Oxford University Press, Oxford (1999)
[29] Borenstein, N., Freed, N.: Multipurpose Internet Mail Extensions. RFC1341, IETF (1992)
[30] Boyes, B.: Systronix White Paper – Why Use Java? Systronix (2010), http://www.PracticalEmbeddedJava.com (last accessed on June 10, 2010)
[31] Brachman, R.: What IS-A Is and Isn't: An Analysis of Taxonomic Links in Semantic Networks. In: Margaria, T. (ed.) Computer, vol. 16(10), pp. 30–36. IEEE Computer Society, Los Alamitos (1983)
[32] Bradshaw, J.M., Sierhuis, M., Acquisti, A., Feltovich, P., Hoffman, R., Jeffers, R., Prescott, D., Suri, N., Uszok, A., Van-Hoof, R.: Adjustable Autonomy and Human-Agent Teamwork in Practice: An Interim Report on Space Applications. In: Hexmoor, H., Facone, R., Castelfranchi, C. (eds.) Agent Autonomy, pp. 243–280. Kluwer, Dordrecht (2002)
[33] Bratman, M.E.: What is Intention? In: Cohen, P.R., Morgan, I.L., Pollock, M.E. (eds.) Communications, pp. 15–32. MIT Press, Cambridge (1990)
[34] Bratman, M.E.: Intentions, Plans and Practical Reason. Center for the Study of Language and Information (1999)
[35] Bréal, M.: Essai de sèmantique. Science des significations, Paris, Hachette (1897); Translated to English by Mrs Henry Cust in 1964
[36] Brooks, R.: A robust layered control system for a mobile robot. IEEE Journal of Robotics and Automation 1(2), 14–23 (1986)
[37] Brown, C.E., Swierenga, S.J., Wellens, A.R.: Social Psychological Metaphors for Human-Computer System Design. In: Proceedings of the IEEE 1991 National Aerospace and Electronics Conference, vol. 2, pp. 793–799. IEEE, Los Alamitos (1991)
[38] Browniee, P.: How the IC took on the World: Part 2. What's New in Electronics 28(8), 7–8 (2009)
[39] Bruneton, E.: ASM 3.0: A Java bytecode engineering library. INRIA, France Telecom (2007)

[40] Brzostek, M.: Gulftown review - months before the launch. PC Lab (2009) (last accessed June 17, 2010)

[41] Buchanan, B.: New research on expert systems. In: Hayes, J., Michied, D. (eds.) Machine Intelligence, vol. 10, pp. 269–299. Wiley, London (1982)

[42] Burnside, W., Neumann, P.M., Mann, A.J.S., Tompson, J.C.: The Collected Papers of William Burnside. Oxford University Press, Oxford (2004)

[43] Burstein, M., Denker, G., Hobbs, J., Kagal, L., Lassila, O.: OWL-S 1.0 Release. In: Martin, D. (ed.) Draft Standard, W3C (2006)

[44] Buyya, R.: Java and its Evolution. In: Powerpoint presentation. Dept. of Computer Science and Software Engineering, The University of Melbourne (2004)

[45] Cahill, V., Gray, E., Seigneur, J.-M., Jensen, C.D., Chen, Y., Shand, B., Dimmock, N., Twigg, A., Bacon, J., English, C., Wagealla, W., Terzis, S., Nixon, P., Serugendo, G.D.M., Bryce, C., Carbone, M., Krukow, K., Nielson, M.: Using Trust for Secure Collaboration in Uncertain Environments. IEEE Pervasive Computing 2(3), 52–61 (2003)

[46] Caldwell, C., Honaker, G.L.: Prime Curios!, Createspace. Scotts Valley, USA (2009)

[47] Callan, R.: Artificial Intelligence. Palgrave MacMillan, England (2003)

[48] Calvert, J.B.: The Slide Rule. Engineering and Technology, 3–5 (2004)

[49] Cardelli, L., Wegner, P.: On Understanding Types, Data Abstraction, and Polymorphism. Computing Surveys 17(4), 471–522 (1985)

[50] Castelfranchi, C.: Guarantees for autonomy in cognitive agent architecture. In: Wooldridge, M., Jennings, N.R. (eds.) Intelligent Agents, ECAI 1994 Workshop on Agent Theories, Architectures, and Languages. LNCS, vol. 890, pp. 56–70. Springer, Netherlands (1977)

[51] Castelfranchi, C., Falcone, R., Pezzulo, G.: Trust in Information Sources as a Source for Trust: A Fuzzy Approach. In: Proceedings of the Second International Joint Conference on Autonomous Agents and Multiagent Systems, pp. 89–96. ACM Press, New York (2003)

[52] Cerrito, P.B.: Choice of antibiotic in open heart surgery. Intelligent Decision Technologies 1(1-2), 63–69 (2007)

[53] Chambers, A.B., Nagel, D.C.: Pilots of the Future: Human or Computer? Communications of the ACM 28(11), 1187–1199 (1985)

[54] Chan, C.: An expert decision support system for monitoring and diagnosis of petroleum production and separation processes. Expert Systems with Applications 29, 127–135 (2005)

[55] Chauhan, B.K., Hanmandlu, M.: Load forecasting using wavelet fuzzy neural network. International Journal of Knowledge-Based and Intelligent Engineering Systems 13(4), 1327–2314 (2010)

[56] Chira, O., Chira, C., Roche, T., Tormey, D., Brennan, A.: An agent-based approach to knowledge management in distributed design. Journal of Intelligent Manufacturing, The Institution of Engineering and Technology 17(6), 737–750 (2006)

[57] Cohen, D.I.A.: Introduction to Computer Theory, 2nd edn. John Wiley & Sons, Inc., New York (1997)

[58] Cohen, M.S.: A Situation Specific Model of Trust to Decision Aids. Cognitive Technologies (2000)

[59] Cohen, N.H.: Gödel numbers: a new approach to structured programming. SIGPLAN Not. 15(4), 70–74 (1980)

[60] Cohen, P., Levesque, H.: Speech acts and rationality. In: Mann, W.C. (ed.) 23rd Annual Meeting on Association for Computational Linguistics, Chicago, Illinois, pp. 49–60. Association for Computational Linguistics, USA (1985)

[61] Cohen, P.R., Levesque, H.J.: Teamwork. Nous, Special Issue on Cognitive Science and Artificial Intelligence 254(4), 487–512 (1991)

[62] Cohen, P.R., Perrault, C.R.: Elements of a plan-based theory of speech acts. In: Huget, M.-P. (ed.) Communication in Multiagent Systems. LNCS (LNAI), vol. 2650, pp. 1–36. Springer, Heidelberg (2003)

[63] Connolly, D.: Overview of SGML Resources. W3C (1995), http://www.w3.org/MarkUp/SGML/

[64] Coram, R.: Boyd: The Fighter Pilot who changed the Art of War. Little Brown and Company, New York (2002)

[65] Corritore, C.L., Wiedenbeck, S., Kracher, B.: The Elements of Online Trust. In: Extended Abstracts on Human Factors in Computing Systems (CHI 2001), pp. 504–505. ACM Press, New York (2001)

[66] Cox, B., Novabilsky, A.: Object-oriented Programming: An Evolutionary Approach, 2nd edn. Addison-Wesley, Reading (1986)

[67] Cravotta, R.: CPU on a chip Goes public. In: EDN, Reed Elsevier Inc., Netherlands (1971)

[68] Crispin, A., Rankov, V.: Evolutionary algorithm for PCB inspection. International Journal of Knowledge-Based and Intelligent Engineering Systems 13(4), 1327–2314 (2009)

[69] Crook, J.F.: A Pencil-and-Paper Algorithmfor Solving Sudoku Puzzles. Notices of the AMS 56(4), 460–468 (2007)

[70] Curts, R., Campbell, D.: Avoiding information overload through the understanding of OODA Loops, A cognitive hierarchy and object oriented analysis and design. In: Proceedings of the 6th International Command and Control Research and Technology Symposium (CCRTS). US Naval Academy, Annapolis (2001)

[71] Dahl, O., Dijkstra, E.W., Hoare, C.A.R.: Structured Programming. Academic Press, London (1972)

[72] Dahl, O.J., Nygaard, K.: SIMULA - An Algol-Based Simulation Language. Communications of the ACM 9(9) (1966)

[73] Dahm, M.: Byte Code Engineering Library (BCEL) - Version 5.2. Apache Software Foundation (2007)

[74] Dasgupta, S.: Computer Architecture: A Modern Synthesis, vol. 1. John Wiley & Sons, Inc., New York (1989)

[75] Davies, C.: Samsung Announces 128GB SATA II SSDs. Slash Gear, Aradius Inc., Scottsdale, AZ, Ewdison. Then and Vincent. Nguyen and Chris. Davies (2008)

[76] Davies, C.: Samsung Announces 256GB SATA II SSDs. Slash Gear, Aradius Inc., Scottsdale, AZ, Ewdison. Then and Vincent. Nguyen and Chris. Davies (2008)

[77] Davies, C.: Seagate Enterprise SSDs in Development: 2TB HDD in 2009. Slash Gear, Aradius Inc., Scottsdale, AZ, Ewdison. Then and Vincent. Nguyen and Chris. Davies (2008)

[78] Davison, A.: Killer Game Programming in Java. O'Reilly Media, Sebastopol (2005)

[79] Dawidowicz, E.: Towards Smart Intelligent Agents in the Command and Control Environment. In: Proceedings of the 2000 Command and Control Research and Technology Symposium (2000)

[80] de Laplante, R.: JAX-RPC client with Maven2. Java Lobby, DZone (2009) (last accessed June 18, 2010)

[81] De Paz, J.F., Rodríguez, S., Bajo, J., Corchado, J.M.: Case-based reasoning as a decision support system for cancer diagnosis: A case study. Int. J. Hybrid Intell. Syst. 6(2), 97–110 (2009)

[82]  de Kleer, J.: An Assumption-based truth management system (TMS). Artificial Intelligence 28(1), 127–162 (1986)

[83]  Delahaye, J.-P.: The Science behind Sudoku, pp. 80–87. Scientific American, New York (2006)

[84]  Dhar, V., Stein, R.: Intelligent decision support methods: the science of knowledge work. Prentice-Hall, Inc., USA (1997)

[85]  d'Inverno, M., Kinny, D., Luck, M., Wooldridge, M.: A Formal Specification of dMARS. Agent Theories, Architectures, and Languages, 155–176 (1997)

[86]  d'Inverno, M., Luck, M.: Understanding agent systems. Springer-Verlag, Inc.,, New York (2001)

[87]  D'Inverno, M., Luck, M., Georgeff, M., Kinny, D., Wooldridge, M.: The dMars Architecture: A Specification of the Distributed Multi-Agent Reasoning System. In: Autonomous Agents aand Multi-Agent Systems. Kluwer Academic Publishers, Dordrecht (2004)

[88]  Dodson, C.S., Johnson, M.K., Schooler, J.W.: The verbal overshadowing effect: why descriptions impair face recognition. Mem. Cognit. 25(2), 129–139 (1997)

[89]  Doyle, J.: A truth maintenance system. Artificial Intelligence 12, 231–272 (1979)

[90]  Dudek, G., Jenkin, M., Milios, E., Wilkes, D.: Taxonomy for Swarm Robots. In: International Conference on Intelligent Robots and Systems 1993, IROS 1993, IEEE, RSJ, vol. 1, pp. 441–447. IEEE/RSJ, IEEE Press, Piscataway, USA (1993)

[91]  Dzindolet, M.T., Linda, H.P.B., Pierce, G.: Encouraging Human Operators to Appropriately Rely on Automated Decision Aids. In: Research, O.H., Directory, E. (eds.) The 6th International Command and Control Research Program (ICCRTS), September 16-20, pp. 16–20. DTIC, Annapolis (2002)

[92]  Eckel, B.: C++ inside and out. Osborne/McGraw-Hill, USA (1993)

[93]  Eckstein, R.: J2SE Enhancements. Sun Developers Network, Redwood Shores (2006)

[94]  Eckstein, R.: More Enhancements in Java SE 6. Sun Developers Network, Redwood Shores (2006)

[95]  Ehlert, P., Rothkrantz, L.: Intelligent Agent In An Adaptive Cockpit Environment. Technical Report DKE01-01, Version 0.2, Delft University of Technology, Netherlands (2001)

[96]  Emden, M.H.V., Kowalski, R.A.: The Semantics of Predicate Logic as a Programming Language. J. ACM 23(4), 733–742 (1976)

[97]  Englander, R.: Java and SOAP. O'Rielly, Sebastopol (2002)

[98]  Evans, R.: Varieties of learning. In: Rabin, E. (ed.) AI Game Programming Wisdom, vol. 2. Charles River Media, Hingham (2002)

[99]  Falcone, R., Pezzulo, G., Castelfranchi, C.: A Fuzzy Approach to a Belief-Based Trust Computation. In: Falcone, R., Barber, S.K., Korba, L., Singh, M.P. (eds.) AAMAS 2002. LNCS (LNAI), vol. 2631, pp. 73–86. Springer, Heidelberg (2003)

[100]  Fan, X., Sun, S., Sun, B., Airy, G., McNeese, M., Yen, J.: Collaborative RPD-Enabled Agents Assisting the Three-Block Challenge in Command and Control in Complex and Urban Terrain. In: 2005 Conference on Behavior Representation in Modeling and Simulation (BRIMS), Universal City, CA, pp. 113–123 (May 2005)

[101]  Fan, X., Sun, S., Yen, J.: On Shared Situation Awareness for Supporting Human Decision-Making Teams. In: 2005 AAAI Spring Symposium on AI Technologies for Homeland Security, Stanford, CA, pp. 17–24 (2005)

[102]  Fan, X., Yen, J., Miller, M.S., Volz, R.A.: The Semantics of MALLET–An Agent Teamwork Encoding Language. Declarative Agent Languages and Technologies II, 69–91 (2005)

[103]  Fasli, M.: Agent technology for e-commerce. John Wiley, Chichester (2007)

[104] Fazlollahi, B., Vahidov, R.: Multi-agent decision support system incorporating fuzzy logic. In: Fuzzy Information Processing Society, NAFIPS, 19th International Conference of the North American, pp. 246–250 (2000)

[105] Feigenbaum, E.A., Buchanan, B., Lederberg, J.: On generality and problem solving: a case study using dendral program. In: Edinburgh, B., Scottland, M., Michie, D. (eds.) Machine Intelligence, vol. 6, pp. 165–190. Edinburgh University Press (1971)

[106] Ferber, J., Gutkecht, O., Michel, F.: MadKit Development Guide (2005), http:// www.madkit.org/madkit/doc/devguide/devguide.html (last accessed: August 25, 2005)

[107] Finin, T., Labrou, Y., Mayfield, J.: KQML as an Agent Communication Language. In: Software Agents, vol. 480, AAAI Press, The MIT Press, Menlo Park (1997)

[108] Finin, T., Weber, J.: Draft. Specification of the KQML Agent Communication Language. DARPA, External Interfaces Working Group (1993)

[109] Finn, A., Kabacinski, K., Drake, S., Mason, K.: Design Challenges for an Autonomous Cooperative of UAVs. In: Information Decision and Control (IDC 2007), pp. 160–169. IEEE Press, New York (2007)

[110] Fischer, K., Müller, J.P., Pischel, M.: Unifying Control in a Layered Agent Architecture. In: IJCAI 1995, Agent Theory, Architecture and Language Workshop 1995, TM-94-05, 27. Deutsches Forschungszentrum für Künstliche Intelligenz GmbH Erwin-Schrödinger Strasse Postfach 208067608 Kaiserslautern Germany (1994)

[111] Foo, Y.P.S., Kobayashi, H.: The VLSI design automation assistant: prototype system, pp. 184–187. IEEE Computer Society, NY (1986)

[112] Fowler, M.: Writing Software Patterns (2006), http://www.martinfowler. com/articles/writingPatterns.html (last Accessed March 26, 2007)

[113] Francik, J., Fabian, P.: Animating Sign Language in the Real Time. In: Applied Informatics, 20th IASTED International Multi-Conference, Innsbruck, Austria, pp. 276–281 (2002)

[114] Francois, F.I., Chatila, R., Alami, R., Robert, F.: PRS: A High Level Supervision and Control Language for Autonomous Mobile Robots. In: Proceedings of the IEEE International Conference on Robotics and Automation, vol. 1, pp. 43–49. IEEE Press, USA (1996)

[115] Frankel, C.B., Bedworth, M.D.: Control, Estimation and Abstraction in Fusion Architectures: Lessons From Human Information Processing. In: Proceedings of the Third International Conference on Information Fusion (FUSION 2000), vol. 1, MOC–3–MOC–10 (2000)

[116] Franklin, S., Graesser, A.: Is it an Agent, or just a Program?: A Taxonomy for Autonomous Agents. In: Proceedings of the Third International Workshop on Agent Theories, Architectures and Languages, Budapest, Hungary, pp. 193–206 (1996)

[117] Freeman, E., Freeman, E.: Head First: Design Patterns. O'Rielly, CA (2004)

[118] Frege, G.: Extensions As Representative Objects In Frege's Logic. In: Erkenntnis, Ruffino, M. (eds.) Humanities, Social Sciences and Law, vol. 52(2), pp. 239–252. Springer, Netherlands (2004)

[119] Frize, M., Yang, L., Walker, R., O'Connor, A.: Conceptual framework of knowledge management for ethical decision-making support in neonatal intensive care. IEEE Transactions on Information Technology in Biomedicine 9, 205–215 (2005)

[120] Gabrielli, A., Gandolfi, E.: A Fast Digital Fuzzy Processor. IEEE Micro 19(1), 68–79 (1999)

[121] Gamma, E., Helm, R., Johnson, R., Vlissides, J.: Design Patterns: Elements of Reusable Object-Oriented Software. Addison-Wesley, USA (1995)

[122] Gaschnig, J.G.: Performance measurement and analysis of certain search algorithms. Ph.D. thesis. Dept. of Computer science, Pittsburgh, PA, USA (1979)

[123] Gat, E.: ESL: a language for supporting robust plan execution in embedded autonomous agents. In: IEEE Proceedings of Aerospace Conference 1997, vol. 1, pp. 319–324 (1997)

[124] Geary, D.: Decorate your Java code. In: O'Shea, A. (ed.) Java World, Infoworld, Framingham (2001)

[125] Gefen, D.: Reflections on the dimensions of trust and trustworthiness among online consumers. SIGMIS Database 33(3), 38–53 (2002)

[126] Genesereth, M.R., Ketchpel, S.P.: Software agents. Communications of the ACM 37(7), 48–53 (1994)

[127] Genesereth, M.R., Nilsson, N.J.: Logical Foundations of Artificial Intelligence. Morgan Kaufmann, San Francisco (1987)

[128] Georgeff, M., Pell, B., Pollack, M., Tambe, M., Wooldridge, M.: The Belief-Desire-Intention Model of Agency. In: Papadimitriou, C., Singh, M.P., Müller, J.P. (eds.) ATAL 1998. LNCS (LNAI), vol. 1555, pp. 1–10. Springer, Heidelberg (1999)

[129] Gherbi, T., Borne, I., Meslati, D.: MDE and Mobile Agents: Another Reflection on the Agent Migration. In: Proceedings of the UKSim 2009: 11th International Conference on Computer Modelling and Simulation, pp. 468–473. IEEE Computer Society, Washington, DC, USA (2009)

[130] Gilder, G.: Metcalfes Law and Legacy. Technical report, Forbes (1993)

[131] Gleick, J.: Chaos Making a New Science. Penguin Books Ltd., Middlesex (1987)

[132] Glover, F.: Future Paths for Integer Programming and Links to. In: Glover, F., Laguna, M. (eds.) Tabu Search. Kluwer, Norwell (1997)

[133] Glover, F., Laguna, M.: Tabu Search (1997)

[134] Goodman, D.: Dynamic HTML: The Definitive Reference, 3rd edn. John Wiley & Sons, Cambridge (2006)

[135] Gosling, J., McGilton, H.: The Java Language Environment: A White Paper. Sun Microsystems, Mountain View, CA (1995)

[136] Graham, S., Davis, D., Simeonov, S., Daniels, G., Brittenham, P., Nakamura, Y., Fremantle, P., Koenig, D., Zentner, C.: Building Web services with Java: making sense of XML, SOAP, WSDL, and UDDI. Developer's Library (2002)

[137] Greeveer, J.: Theory and examples of Point-Set Topoloy, Belmont, CA (1967)

[138] Grevier, D.: AI – The Tumultuous History of the Search for Artificial Intelligence. Basic Books, New York (1993)

[139] Grosz, B., Kraus, S.: Collaborative Plans for Complex Group Activity. In: Leake, D. (ed.) Artificial Intelligence, vol. 86(2), pp. 269–357. AAAI, Menlo Park (1996)

[140] Gruber, T.R.: A Translation Approach to Portable Ontology Specifications. Knowledge Acquisition 5(2), 199–220 (1993)

[141] Guha, R.K.: Dynamic microprogramming in a time sharing environment. In: MICRO 2010: Proceedings of the 10th Annual Workshop on Microprogramming, pp. 55–60. IEEE Press, USA (1977)

[142] Gunter, C.A.: Semantics of Programming Lanuages: Structure and Techniques. MIT Press, Cambridge (1992)

[143] Guo, S., Peters, L., Surmann, H.: Design and Application of an Analog Fuzzy Logic Controller. IEEE Transactions on Fuzzy Systems 4(4), 429–438 (1996)

[144] Hammond, G.T.: The Mind of War: John Boyd and American Security. Smithsonian Institution Press, Washington, USA (2004)

[145] Hanratty, T., Dumer, J., Yen, J., Fan, X.: Using Agents with Shared Mental Model to Support Objective Force. In: Cybernetics and Informatics (SCI 2003), 7th World Multi Conference on Systemics, Orlando, USA, pp. 27–30 (July 2003)

[146] Hansen, M.: SOA Blog. Service Centric, USA (2006)

[147] Harper, L.W., Delugach, H.S.: Using Conceptual Graphs to Represent Agent Semantic Constituents. In: Ganter, B. (ed.) ICCS 2004. LNCS (LNAI), vol. 3127, pp. 333–345. Springer, Heidelberg (2004)

[148] Hayes, B.: Unwed Numbers. American Scientist 94(1), 12–15 (2006)

[149] Heaton, J.: Programming Spiders, Bots, and Aggregators in Java. Sybex, San Fransisco (2002)

[150] Heinze, C., Goss, S., Josefsonn, T., Bennett, K., Waugh, S., Lloyd, I., Murray, G., Oldfield, J.: Interchanging Agents and Humans in Military Simulation. AI Magazine 23(2), 37–48 (2002)

[151] Hennessy, J.L., Patterson, D.A.: Computer Architecture: A Quantitative Approach, 4th edn. Elsevier Science Inc., Amsterdam (2007)

[152] Henry, K.: JDBC - Java Database Connectivity. ACM, Inc., New York (2001)

[153] Henz, M., Troung, H.-M.: SudokuSAT - A Tool for analyzing dificult Sudoku. In: Kacprzyk, J. (ed.) Tools and Applications with Artificial Intelligence. SCI, vol. 166, pp. 23–35. Springer Berlin, Heidelberg (2009)

[154] Higman, B.: A comparative study of Programming Languages, vol. 48. MacDonald, London (1967)

[155] Hoare, C.: Hints on Programming Language Design. In: Wasserman, A.I. (ed.) Tutorial Programming Language Design, pp. 43–52. IEEE Computer Society, Los Alamitos (1980)

[156] Hoffman, R.R.: Whom (or What) Do You (Mis)Trust?: Historical Reflections on the Psychology and Sociology of Information Technology. In: Proceedings of the Fourth Annual Symposium on Human Interaction with Complex Systems, pp. 28–36 (1998)

[157] Hofstede, G.: Culture's consequences: international differences in work related values. Sage, CA (1980)

[158] Hofstede, G.: Cultures and Organisations: Software of the mind. McGraaw-Hill ,Maidenhaid, England (1991)

[159] Holt, C.K.: Sudoku Puzzle: An Exercise in Constraint Programming and Visual Prolog. In: The First Visual Prolog Applications and Language Conference, Portugul, April 24-26, pp. 84–89. Prolog Development Centre, Denmark (2006)

[160] Hopkins, M., DuBois, C.: New software can help people make better decisions in time-stressed situations. Science Daily, 1–2 (2005)

[161] Howden, N., Rönnquist, R., Hodgson, A., Lucas, A.: JACK Intelligent Agents: Summary of an Agent Infrastructure. In: 5th International Conference on Autonomous Agents 2001, Montreal, Canada (2001)

[162] Hritcu, C., Buraga, S.: A reference implementation of ADF (Agent Developing Framework): semantic Web-based agent communication. In: Seventh International Symposium on Symbolic and Numerical Algorithms for Scientific Computing, Timisoara, Romania, vol. 8 (2006)

[163] Hudson, P.: The Industrial Revolution. Oxford University Press, USA (1992)

[164] Huhns, M.N., Singh, M.P., Ksiezyk, T.: Global Information Management via Local Autonomous Agents. In: In Proceedings of the 13th International Workshop on Distributed Artificial Intelligence, Seattle, WA, pp. 36–45. Morgan Kaufmann, San Francisco (1994)

[165] Hunt, G.: Mission Planning Systems for Tactical Aircraft (Preflight and In Flight). Technical Report 313, AGARD Joint Working Group 15, Neuillysurseine, France (1992)

[166] Ichalkaranje, N., Sioutis, C., Urlings, P., Tweedale, J., Lui, F., Jain, L.: Intelligent Agents for Airborne Mission Systems and Land Operations Environments: Simulation Environment Report. Research Report EIE-KES-IAAMS-2002-LJNI-07 [WP9], University of South Australia, Mawson Lakes, Adelaide (December 2003)

[167] Iliadis, L.: A decision support system applying an integrated fuzzy model for long-term forest fire risk estimation. Environmental Modelling and Software 20, 613–621 (2005)

[168] ISO:IEC. Information Technology - Basic Reference Model. In: ISO 7498-1. International Standards Organisation, Switzerland (1982)

[169] ISO/IEC. Information Technology - Basic Reference Model. In: ISO 7498-1. International Standards Organisation, Switzerland (1982)

[170] Jain, A., Jain, A., Jain, S., Jain, L. (eds.): Artificial Intelligence Techniques in Breast Cancer Diagnosis and Prognosis, Machine Perception and Artificial Intelligence, vol. 39. World Scientific Publishing, USA (2000)

[171] Jain, L.C., Jain, R.K.: Hybrid Intelligent Engineering Systems. World Scientific Publishing Company, Singapore (1997)

[172] Jain, L.C., Johnson, R.P., Takefuji, Y., Zadeh, L.A. (eds.): Computational Intelligence Techniques in Industry. CRC Press, USA (1998)

[173] Jain, L.C., Vemuri, R. (eds.): Industrial Applications of Neural Networks. CRC Press, USA (1988)

[174] Jarvis, J.: JACK Intelligent Agents: Teams Manual 3.0. In: Agent Oriented Systems, Melbourne (2004)

[175] Jennings, N., Wooldridge, M.: Software Agents, vol. 42(1), pp. 17–20. IEEE Press, N.Y (1996)

[176] Jennings, N.R., Wooldridge, M.: Applications of intelligent agents. In: Agent Technology: Foundations, Applications, and Markets, pp. 3–28. Springer-Verlag, Inc., USA (1998)

[177] Jiang, E.: Detecting spam email by radial basis function networks. International Journal of Knowledge-Based and Intelligent Engineering Systems 11(6), 409–418 (2007)

[178] Job, D., Shankararaman, V., Miller, J.: Combining CBR and GA for designing FPGAs. In: Proceedings of Third International Conference on Computational Intelligence and Multimedia Applications, ICCIMA 1999, pp. 133–137 (1999)

[179] Johnson, P.E.: The Genesis and development of Set Theory. The Two-Year College Mathematical Journal 3(1), 55–62 (1972)

[180] Johnson, R.E.: Frameworks = (components + patterns). Commun. ACM 40(10), 39–42 (1997)

[181] Johnston, N.: CRM: Cross-Cultural Perspectives. In: Wiener, E.L., Kanki, B.G., Helmreich, R.L. (eds.) Cockpit Resource Management, vol. 13, pp. 367–398. Academic Press, San Diego (1993)

[182] Jones, B.: Sleepers, Wake!: Technology and the Future of Work, 2nd edn. Oxford University Press, Inc., New York (1990)

[183] Kanoh, H.: Dynamic route planning for car navigation systems using virus genetic algorithms. International Journal of Knowledge-Based and Intelligent Engineering Systems 11(1), 65–78 (2007)

[184] Karnaugh, M.: The Map Method for Synthesis of Combinatorial Logic Circuits. Transformations of American Institute of Electrical Engineers 72(9), 593–599 (1952)

[185] Kay, A.C.: The Early History of Smalltalk, 28th edn., pp. 69–95. ACM, Inc., New York (1993)

[186] Kayten, P.J.: The Accident Investigatorís Perspective. In: Wiener, E.L., Kanki, B.G., Helmreich, R.L. (eds.) Cockpit Resource Management, vol. 10, pp. 283–314. Academic Press, San Diego (1993)

[187] Keene, S.G.: Object-Oriented Programming in Common Lisp: A Guide to CLOS. Addison-Wesley, Massachusetts (1989)

[188] Kelly, C., Boardman, M., Goillau, P., Jeannot, E.: Guidelines for Trust in Future ATM Systems: A Literature Review. Technical Report 030317-01, European Organisation for the Safety of Air Navigation (May 2003)

[189] Bowen, K.A., Buettner, K.A., Cicekli, L., Turk, A.K.: The design and implementation of a high-speed incremental portable Prolog compiler. In: Shapiro, E. (ed.) ICLP 1986. LNCS, vol. 225, pp. 650–656. Springer, Heidelberg (1986)

[190] Kerievsky, J.: The Checks pattern language of information integrity. In: Patern Languages of Program Design, Portland, Oregon, Cunningham (1994)

[191] Kernighan, B.W., Ritchie, D.M.: The C Programming Language. Prentice-Hall, Englewood Cliffs (1978)

[192] Khasawneh, M.T., Bowling, S.R., Jiang, X., Gramopadhye, A.K., Melloy, B.J.: A Model for Predicting Human Trust in Automated Systems. In: Proceedings of the Eigth Annual International Conference of Industrial Engineering - Theory, Applications and Practice, Las Vegas, Nevada, USA, pp. 216–222 (2003)

[193] Khazab, M., Tweedale, J., Jain, L.: Web-based multi-agent system architecture in a dynamic environment. International Journal of Knowledge-Based and Intelligent Engineering Systems 14(4), 217–227 (2010)

[194] Kim, I.-G., Hong, J.-E., Bae, D.-H., Han, I.-J., Youn, C.: Scalable Mobile Agents Supporting Dynamic Composition of Functionality. In: Wagner, T.A., Rana, O.F. (eds.) AA-WS 2000. LNCS (LNAI), vol. 1887, pp. 199–213. Springer, Heidelberg (2001)

[195] Kinny, D., Georgeff, M., Rao, A.: A Methodology and Modelling Techniques for Systems of BDI Agents. In: Perram, J., Van de Velde, W. (eds.) MAAMAW 1996. LNCS, vol. 1038, pp. 56–71. Springer, Heidelberg (1996)

[196] Klein, G., Calderwood, R., MacGregor, D.: Critical decision method of eliciting knowledge. IEEE Transactions on Systems, Man and Cybernetics 19, 462–472 (1989)

[197] Klein, G.A., Crandell, B.W.: The role of simulation in problem solving and decision making. In: Hancock, P., Flach, J., Caird, J., Vincente, K. (eds.) Local Applications of the Ecological Approach to Human-Machine Systems, pp. 47–92. Lawrence Erlbaum Associates, London (1995)

[198] Klien, G.: Sources of Power. MIT Press, Camabridge (1998)

[199] Klien, G.A.: Recognition-primed decisions. In: Rouse, W.B. (ed.) Advances in Man Machine System Research, vol. 5, pp. 47–92. JAI Press, USA (1989)

[200] Klien, G.A.: A Recognition-Primed Decision (RPD) model of rapid decision making. In: Klien, G.A., Orasanu, J., Calderwood, R., Zsambok, C.E. (eds.) Decision Making in Action: Models and Methods, pp. 138–147. Aplex Publishing Coroporation, New Jersey (1993)

[201] Knuth, D.: The Art of Computer Programming. In: Grace Murray Hopper Award. Association for Computing Machinery. ACM Press, New York (1971)

[202] Koehl, S.: What would you do with 80 cores? In: Research@Intel,Santa Clara, USA, pp. 1–6(2007)

[203] Kotsiantis, S., Tzelepis, D., Koumanakos, E., Tampakas, V.: Selective costing voting for bankruptcy prediction. International Journal of Knowledge-Based and Intelligent Engineering Systems 11(2), 409–418 (2007)

[204] Kubota, N., Kamijima, S.: Intelligent control for vision-based soccer robots. International Journal of Knowledge-Based and Intelligent Engineering Systems 10(1), 83–91 (2007)

[205] Kuru, S., Caolayan, M.: Crew Assistant: AIP survey report. Technical Report DAIS-RWP3.4BU, EUCLID RTP 6.5 Crew Assistant, Bogaziç Unversity, Istanbul, Turkey (1995)

[206] Lab, T.I.: JADE - Java Agent Development Framework (2007)

[207] Labrou, Y., Finin, T.: Semantics for an Agent Communication Language. In: Agent Theories, Architectures, and Languages, 4th International Workshop, ATAL Proceedings, Intelligent Agents IV, pp. 209–314. Springer, USA (1998)

[208] Labrou, Y., Finin, T., Peng, Y.: Agent Communication Languages: The Current Landscape. IEEE Intelligent Systems and Their Applications 14(2), 45–52 (1999)

[209] Labrou, Y., Finin, T., Peng, Y.: The current Landscape in Agent Communication Languages. IEEE Intelligent Systems 14(2), 1–11 (1999)

[210] Laird, J., Newell, A., Rosenbloom, P.: SOAR: architecture for general intelligence. Artificial Intelligence 33(1), 1–64 (1987)

[211] Lambert, D., Scholz, J.: Ubiquitous command and control. Intelligent Decision Technologies 1(3), 157–173 (2007)

[212] Lange, D.B.: Java Aglet Application Programming Interface (J-AAPI) White Paper - Draft 2. Technical report, IBM Tokyo Research Laboratory (1997)

[213] Lange, D.B.: Mobile agents: environments, technologies, and applications. In: Third International Conference and Exhibition on The Practical Application of Intelligent Agents and Multi-Agent Technology (PAAM 1998), pp. 11–14. The Practical Application Company, London (1998)

[214] Lange, D.B., Oshima, M.: Mobile Agents with Java: The Aglet API. World Wide Web 1(3), 111–121 (1998)

[215] Lange, D.B., Oshima, M., Karjoth, G., Kosaka, K.: Aglets: Programming Mobile Agents in Java. In: Masuda, T., Tsukamoto, M., Masunaga, Y. (eds.) WWCA 1997. LNCS, vol. 1274, pp. 253–266. Springer, Heidelberg (1997)

[216] Lauralee, S.: Human Physiology - From cells to systems, 7th edn. Brooks - Cole, CA (2007)

[217] Lazzaro, J., Ryckebusch, S., Mahowald, M.A., Mead, C.A.: Winner-Take-All Networks of O(N) Complexity. Report CS-TR-88-21, Caltech, Pasadena, CA, USA (1988)

[218] Leach, P.J., Salz, R.: Transmission Control Protocol/Internet Protocol - RFC 793. In: IETF. Networking Division of the USC Information Sciences Institute (ISI), Marina del Rey (1997)

[219] Leedy, D., Ormrod, J.: Practical Research: Planning and Design, 8th edn. Person Press, New Jersey (2001)

[220] Leibniz, G.W.: Explication de l'Arithmétique Binaire. Memoires de l'Académie Royale des Sciences 3, 85–89 (1903)

[221] Lemay, L., Perkins, C.: Teach Yourself Java in 21 Days. Sams, Indianapolis (1996)

[222] Levesque, H.J., Reiter, R., Lesperance, Y., Lin, F., Scherl, R.B.: GOLOG: A Logic Programming Language for Dynamic Domains. Journal of Logic Programming 31(1-3), 59–83 (1997)

[223] Lewis, B., Berg, D.J.: Threads primer: a guide to multithreaded programming. Prentice Hall Press, Upper Saddle River (1995)

[224] Lewis, R.: Metaheuristics can solve sudoku puzzles. Journal of Heuristics 13(4), 387–401 (2007)

[225] Lewis, T. (ed.): Object oriented application frameworks. Manning Publications Co., USA (1995)

[226] Listing, J.B.: Introductory Studies in Topology. Vandenheock und Represcht, Gottingen (1847)

[227] Liu, T.-I., Khan, W.H., Oh, C.T.: A knowledge-based system of high speed machining for the manufacturing of products. International Journal of Knowledge-Based and Intelligent Engineering Systems 14(4), 185–199 (2010)

[228] Luck, M., Ashri, R., D'Inverno, M.: Agent-Based Software Development. Artech House, London (2004)

[229] Luger, G.F.: Artificial Intelligence, 5th edn. Addison-Wesley, Reading (2004)

[230] Lund, T., Torralba, A., Carvajal, R.: The architecture of an FPGA-style programmable fuzzy logic controller chip. In: 5th Australasian Computer Architecture Conference, ACAC 2000, pp. 51–56 (2000)

[231] Mackworth, A.: The Coevolution of AI and AAAI. AI Magazine 26, 51–52 (2005)

[232] Madsen, M., Gregor, S.: Measuring human-computer trust. In: Proceedings of the Eleventh Australasian Conference on Information Systems, Brisbane (2000)

[233] Maes, P.: The agent network architecture ANA. SIGART Bulletin 2(4), 115–120 (1991)

[234] Maes, P.: Agents that Reduce Work and Information Overload. Communications of the ACM 37(7) (1994)

[235] Magazine, A.D.: RPDE delivers Rapid Progress (2007)

[236] Mailer, G.: A Guess-Free Sudoku Solver. Technical Report COM3050. University of Sheffield, UK (2008)

[237] Mankala, M.R., Dharmaraj, S. (eds.): JavaServer Pages Technology. Sun Microsystems, Inc, Menlo Park (2006)

[238] Mann, R.: Interpersonal styles and group development. American Journal of Psychology 81, 137–140 (1970)

[239] Martin, D.: OWL-S: Semantic Markup for Web Services. DAML-S Organisation (1993)

[240] Martinez, P., Garcia-Serrano, A.M., Calle-Gomez, J., Rodrigo-Aguado, L.: An Agent-based Design of a NL-Interaction for Intelligent Assistance in an E-Commerce Scenario. In: Preceedings of the IEEE International Conference on Systems, Man, and Cybernetics (IEEE SMC 2001), vol. 1, pp. 530–535 (2001)

[241] Mayk, I., Yaeger, G., Lossau, K., Langston, J.: A Knowledge Based Doctrine Tool For Command And Control. In: Command & Control Research & Technology Symposium, IOS Press, Amsterdam (1998)

[242] McAllester, D.: An outlook on truth maintenance. Technical Report 551, AI Laboratory, MIT (1980)

[243] McCarthy, J.: Programs with common sense. In: Symposium on Mechanization of Thought Processes. National Physical Laboratory, England (1958)

[244] McCarthy, J., Abrahams, P.W., Edwards, D.J., Hart, T.P., Levin, M.I.: LISP 1.5 Programmer's Manual, 2nd edn. MIT Press, Boston (1962)

[245] McCarthy, T.: The sayings of John McCarthy. Stanford University, California (1998)

[246] McCorduck, P.: Machines who think. Freeman, pp. 1–375 (1979)

[247] McCulloch, W.S., Pitts, W.H.: A logical calculus of the ideas immanent in nervous activity. Bulletin of Mathematical Biophysics 5, 115–133 (1943)

[248] McGloughlin, S.: Multimedia Quotes: Concepts and Practice. Prentice-Hall, Upper Saddle River (2001)

[249] McIlroy, M.D.: Mass Produced Software Components. In: NATO conference on Software Engineering, Brussels, pp. 138–155 (1968)

[250] McLaughlin, B., Flanagan, D.: Java 5.0 Tiger: A Developer's Notebook. O'Reilly, Cambridge (2004)

[251] Mepham, M.: Solving Sudoku, Crosswords. Frome, Somerset (2005)

[252] Messerschmidtt, D.G.: Networked Application: A guide to the New Computing Infrastructure. Morgan Kaufmann, San Francisco (1999)

[253] Miller, A.: What to expect in Java SE 7. Java World 12, 1–8 (2008)

[254] Minsky, M.: Society of Mind. Simon and Schuster, Pymble, Australia (1985)

[255] Minsky, M.L., Papert, S.A.: Perceptrons. MIT Press, Cambridge (1969)

[256] Moor, J.H.: Briefly noted: the turing test: The elusive standard of artificial intelligence. Comput. Linguist. 30(1), 115–116 (2004)

[257] Moore, G.E.: Cramming more components onto integrated circuits. Electronics 38(8), 1–4 (1965)

[258] Moray, N.: Intelligent aids, mental models, and the theory of machines. Int. J. Man-Mach. Stud. 27(5-6), 619–629 (1987)

[259] Morris, J.: AMD's Extreme Makeover What the new roadmaps reveal. The Core Truth, ZDNet (2008) (last Accessed June 17, 2010)

[260] Muller, J., Pischel, M.: The Agent Architecture InteRRaP: Concept and Application. Technical Report RR-93-26, DFKI, Saarbrucke (1993)

[261] Mulley, G.P.C.: A design tool for performance prediction of real-time systems. Ph.D. thesis, University of Reading, UK (1986)

[262] Muscettola, N., Nayak, P.P., Pell, B., Williams, B.C.: Remote Agent: To Boldly Go Where No AI System Has Gone Before. Artificial Intelligence 103(1-2), 5–47 (1998)

[263] Nadathur, G., Miller, D.: Higher-order Horn clauses. J. ACM 37(4), 777–814 (1990)

[264] Naftalin, M., Wadler, P.: Java Generics and Collections. O'Rielly, CA (2006)

[265] Nareyek, A. (ed.): Constraint-Based Agents Modeling and Local-Search-Based Reasoning for Planning and Schedulling in Open and Dynamic Worlds. LNCS (LNAI), vol. 2062. Springer, Heidelberg (2001)

[266] Nayak, U., Williams, B.C.: Fast context switching in real-time propositional reasoning. In: Proceedings of AAAI 1997, pp. 50–56 (1997)

[267] Negnevitsky, M.: Artificial Intelligence, A Guide to Intelligent Systems. Pearson Education Limited, London (2005)

[268] Negoita, M.G., Arslan, T.: Adaptive hardware/evolvable hardware – The state of the art and the prospectus for future development. International Journal of Knowledge-Based and Intelligent Engineering Systems 12(3), 183–185 (2009)

[269] Newell, A.: Production Systems: Models of Control Structure. In: Chase, W.G. (ed.) Visual and Information Processing, pp. 463–526. Academic Press, San Diego (1973)

[270] Newell, A.: Physical symbol systems. Cognitive Science: A Multidisciplinary Journal 4(2), 135–183 (1980)

[271] Newell, A., Card, S.: Straightening Out Softening Up: Response to Carroll and Campbell. Human-Computer Interaction 2(3), 251–267 (1986)

[272] Newell, A., Card, S.K.: The Prospects for Psychological Science in Human-Computer Interaction. Human-Computer Interaction 1(3), 209–242 (1985)

[273] Newell, A., Simon, H.A.: GPS, a program that simulates human thought. In: Feigenbaum, E.A., Feldman, J. (eds.) Computers and Thought, pp. 279–296. McGraw-Hill, New York (1963)

[274] Newell, A., Simon, H.A.: Human Problem Solving. Prentice-Hall, Englewood Cliffs (1972)

[275] Newell, A., Simon, H.A.: GPS, a program that simulates human thought. Computation & intelligence: collected readings, 415–428 (1995)

[276] Newell, A., Simon, H.A.: Computer Science as Empirical Inquiry: Symbols and Search. Communications of the ACM 19(3), 113–126 (1976)

[277] Nielsen, J. (ed.): Advances in Human/Computer Interaction. Ablex Publishing Corporation, Greenwich (1995)

[278] Nii, H.P.: Blackboard Systems: the Blackboard Model of Problem Solving and the Evolution of Blackboard Architectures. AI Magazine 7(2), 82–107 (1986)

[279] Niklaus, W.: On the Design of Programming Languages. In: Information Processing, pp. 386–393. North Holland Publishing Co., New York (1974)

[280] Nwana, H.S.: Software Agents: An Overview. In: McBurney, P. (ed.) The Knowledge Engineering Review, vol. 11(3), pp. 205–244. Simon Parsons, USA (1996)

[281] Nwana, H., Wooldridge, M.: Software agent technologies. In: Nwana, H.S., Azarmi, N. (eds.) Software Agents and Soft Computing: Towards Enhancing Machine Intelligence. LNCS, vol. 1198, pp. 59–78. Springer, Heidelberg (1997)

[282] Nwana, H.S., Ndumu, D.T., Lee, L.: ZUES: An advanced tool-kit for engineering distributed multi-agent systems. Applied AI 13(1(2)), 129–1185 (1998)

[283] Odell, J. (ed.): Foundation of Intelligent Physical Agents. IEEE Computer Society Press, New York (2005)

[284] O'Keefe, R.A.: The craft of Prolog. Logic Programming (1990)

[285] Oliva, E., Viroli, M., Omicini, A.: Minority Game: A Logic-Based Approach in TuC-SoN. In: Paoli, F.D., Stefano, A.D., Omicini, A., Santoro, C. (eds.) 7th WOA 2006 Workshop, From Objects to Agents, Catania, Italy (September 26-27, 2006)

[286] O'Regan, G.: Introduction to Aspect-Oriented Programming. O'Reilly, Sebastopol (2004)

[287] Overman, E.S.: The New Sciences of Administration: Chaos and Quantum Theory. Public Administration Review 556 (1996)

[288] Panait, L., Luke, S.: Cooperative Multi-Agent learning: The state of the art. Autonomous Agents and Multi-Agent Systems 11(3), 387–434 (2005)

[289] Pearl, J.: Heuristics: intelligent search strategies for computer problem solving. Addison-Wesley Longman Publishing Co., Inc., Boston (1984)

[290] Pegg, E.: Math Games: Sudoku Variations. Mathematical Association of America, Washington D.C (2005)

[291] Pemberton, S., Austin, D., Axelsson, J., Çelik, T., Dominiak, D., Elenbaas, H., Epperson, B., Ishikawa, M., Matsui, S., McCarron, S., Navarro, A., Peruvemba, S., Relyea, R., Schnitzenbaumer, S., Stark, P.: XHTML 1.0 The Extensible HyperText Markup Language. W3C (2002), http://www.w3.org/TR/xhtml1/

[292] Perez, M., Marwala, T.: Stochastic Optimization Approaches for Solving Sudoku. In: International Workshop on Stocastic and Applied Global Optimization, vol. 0805-0697, p. 697. University of Witwatersand, Johannesburg (2008)

[293] Phillips-Wren, G., Jain, L.: Intelligent decision support systems in agent-mediated environments. In: Phillips-Wren, G., Jain, L. (eds.) Frontiers in Articial Intelligence and Applications, vol. 115, pp. vii – ix. IOS Press, Amsterdam (2005)

[294] Pollard, C., Sag, I.A.: An Information-based Approach to Syntax and Semantics: Fundamentals. Centre for the study of Language and Information (CSIL), vol. 13. Stanford University Press, Chicago (1987)

[295] Postel, J.: User Datagram Protocol - RFC 768. IETF. Networking Division of the USC Information Sciences Institute (ISI), Marina del Rey (1980)

[296] Postel, J.: Transmission Control Protocol/Internet Protocol - RFC 793. IETF. Networking Division of the USC Information Sciences Institute (ISI), Marina del Rey (1981)

[297] Postel, J., Reynolds, J.: File Transfer Protocol - RFC 959. IETF. Networking Division of the USC Information Sciences Institute (ISI), Marina del Rey (1985)

[298] Power, K.: Object-Oriented Disciplines. Informatics 1(4), 35–38 (1993)

[299] Prince, C., Salas, E.: Training and Research for Teamwork in the Military Aircrew. In: Wiener, E., Kanki, B., Helmreich, R. (eds.) Cockpit Resource Management, ch. 12, pp. 337–364. Academic Press, San Diego (1993)

[300] Prinzel, L.J.: The Relationship of Self-Efficacy and Complancency in Pilot-Automation Interaction. Technical Report TM-2002-211925, NASA, Langley Research Center, Hampton, Virginia (2002)

[301] Quteishat, A., Lim, C.P., Tweedale, J., Jain, L.C.: A Multi-Agent Classifier System based on the TNC Model. In: 12th Online World Conference on Soft Computing in Industrial Applications. World Federation of Soft Computing (2007)

[302] Raggett, D.: HTML 4.01 Specification. W3C (1999), http://www.w3.org/TR/REC-html40/

[303] Rao, A., Georgeff, M.: BDI Agents: From Theory to Practice. In: 1st International Conference on Multi-Agent Systems (ICMAS 1995), San Francisco, CA, pp. 312–319 (1995)

[304] Rasmussen, J.: Outlines of a hybrid model of the process plant operator. In: Sheridan, T.B., Johannsen, G. (eds.) Monitoring Behaviour and Supervisory Control. Plenum Press, New York (1976)

[305] Rasmussen, J.: Information Processing and Human-Machine Interaction: An Approach to Cognitive Engineering. Elsevier Science Inc., Amsterdam (1986)

[306] Rasmussen, J.: Skills, rules, and knowledge; signals, signs, and symbols, and other distinctions in human performance models. In: System Design for Human Interaction, pp. 291–300. IEEE Press, USA (1987)

[307] Rasmussen, J.: Diagnostic Reasoning in Action. IEEE Transactions on Systems, Man, and Cybernetics 23(4), 981–992 (1993)

[308] Rasmussen, J., Pejtersen, A.M., Goodstein, L.P.: Cognitive Systems Engineering. John Wiley & Sons, Inc., Chichester (1994)

[309] Rasmussen, J., Pejtersen, A.M., Schmidt, K.: Taxonomy for Cognitive Work Analysis. Technical report, Risø National Laboratory (September 1990)

[310] Rau, B.R., Glaeser, C.D., Greenawalt, E.M.: Architectural support for the efficient generation of code for horizontal architectures. SIGARCH Comput. Archit. 10(2), 96–99 (1982)

[311] Ren, X., Thompson, H.A., Fleming, P.J.: An agent-based system for distributed fault diagnosis. KES Journal 10(5), 319–335 (2006)

[312] Reynolds, C.W.: Flocks, Herds, and Schools: A Distributed Behavioral Model. Computer Graphics 21(4), 25–34 (1987)

[313] Rich, E., Knight, K.: Artificial Intelligence. McGraw-Hill, New York (1991)

[314] Richards, D., van Splunter, S., Sabou, M.: An Experience report on using DAML-S. In: Workshop on E-Services and the Semantic Web (ESSW 2003), The Twelfth International World Wide Web Conference, Budapest, Hungary (2003)

[315] Roetter, A.: Writing multithreaded Java applications, Teton Data Systems, Jackson, WY. IBM Developer (2001)

[316] Roman, R., Sriganesh, P., Brose, G.: Mastering Enterprise JavaBeans. Wiley, New York (2005)

[317] Rooij, A.V., Jain, L.C., Johnson, R.P.: Neural Network Training Using Genetic Algorithms. World Scientific Publishing Company, Singapore (1996)

[318] Rosenbloom, A.: Trusting Technology: Introduction. Communications of the ACM 43(12), 31–32 (2000)

[319] Rudowsky, I.: Intelligent Agents. Communications of the Association for Information Systems, 275–190 (2004)

[320] Russel, S., Norvig, P.: Artificial Intelligence: A Modern Approach. Prentice Hall series in Artificial Intelligence, 2nd edn. Prentice-Hall, Englewood Cliffs (2003)

[321] Ryan, P., Zalcman, L.: The DIS vs HLA Debate: What's in it for Australia? In: SimTect 2003, pp. 1–6. Simulation Industry Associatioin of Australia (2003)

[322] Saaresto, M.: A Review On Java API For XML Messaging (JAXM). Technical report, Draft SOAP Standard, W3C (2000)

[323] Sammet, J.: Programming Languages: History and Fundamentals. Prentice-Hall, Englewood Cliffs (1969)

[324] Samuels, A.: Studies in Machine Learning Using the Game of Checkers. IBM Journal of Research & Development 3, 211–229 (1959)

[325] Satchell, P.M.: Cockpit Monitoring and Alerting Systems. Ashgate Publishing Ltd., Hampshire (1993)

[326] Sato, M., Sato, Y., Jain, L.C.: Fuzzy Clustering Models and Applications. Springer, Heidelberg (1997)

[327] Schmidt, D.C.: Patterns for Concurrent, Parallel, and Distributed Systems. Washington State University, St. Louis, School of Engineering & Applied Science, Missouri (2005)

[328] Schneidewind, N.F.: Reliability Modeling for Safety-critical Software. IEEE Transactions on Reliability 46(1), 88–98 (1997)

[329] Schütz, H.: Generating Minimal Herbrand Models Step by Step. In: Murray, N.V. (ed.) TABLEAUX 1999. LNCS (LNAI), vol. 1617, pp. 263–277. Springer, Heidelberg (1999)

[330] Searle, J.R.: Speech Acts. Cambridge University Press, Cambridge (1969)

[331] Seely, S., Sharkey, K.: SOAP: Cross Platform Web Services Development Using XML. Pearson Education, London (2001)

[332] Shilov, A.: Playstation 3 Launches with BE CPU. X-Bit Laboratories, USA (2006)

[333] Shirazi, J.: Hotpatching a Java 6 Application. Fasterj, UK (2010)

[334] Shneiderman, B.: Designing trust into online experiences. Communications of the ACM 43(12), 57–59 (2000)

[335] Shoham, Y.: An overview of agent-oriented programming. In: Bradshaw, J.M. (ed.) Software Agents, vol. 4. AAAI Press, Menlo Park (1997)

[336] Shortliffe, E.: MYCIN: Computer-Based Medical Consultations. Elsevier Press, New York (1976)

[337] Simmons, R.: Generate, test and debug: a paradigm for combining associational and causal reasoning. In: Krivine, D., Simmons, R. (eds.) Second Generation Expert Systems, pp. 79–92. Springer, Heidelberg (1993)

[338] Simonis, H.: Sudoku as a constraint problem. Imperial College of Science, Technology, and Medicine, Center for Planning and Resource Control, 13–27 (2005)

[339] Singh, M.P.: Agent Communication Languages: Rethinking the Principles. Carolina State Uni. (1998)

[340] Singleton, A.: Wired on the Web. In: Byte, vol. 21(1), pp. 77–80. McGraw-Hill, London (1996)

[341] Smith, D.: So you thought Sudoku came from the land of the rising sun. The Observer (2005)

[342] Smith, D.R.: KIDS: a semiautomatic program development system. IEEE Transactions on Software Engineering 16, 1024–1043 (1990)

[343] Smith, I.A., Smith, I.A., Cohen, P.R., Bradshaw, J.M., Greaves, M.A.G.M., Holmback, H.A.H.H.: Designing conversation policies using joint intention theory. In: Cohen, P.R. (ed.) Proceedings of International Conference on Multi Agent Systems, 1998, pp. 269–276 (1998)

[344] Sodabot, M.H.C.: A software agent construction system. Technical report. MIT AI Laboratory, Cambridge (1995)

[345] Soropika, D.: Soboran: Japanese Abacus, pp. 1–2. The League of Japan Abacus Associations, Tokyo (2004)

[346] Sperberg-McQueen, C.M., Thompson, H.: XML Schema (2000),
http://www.w3.org/XML/Schema

[347] Staker, R.: Use of Bayesian belief networks in the analysis of information system network risk. In: Proceedings of Information, Decision and Control, IDC 1999, pp. 145–150 (1999)

[348] Steels, L.: Adaption Multi-Agent Learning. In: Alonso, E., Kudenko, D., Kazakov, D. (eds.) AAMAS 2003. LNCS (LNAI), vol. 2636, pp. 559–572. Springer, Heidelberg (2003)

[349] Sterling, L., Shapiro, E.: The Art of Prolog: Advanced Programming Techniques. In: Logic Programming. MIT Press, Cambridge (1994)

[350] Stevens, J.: The How and Why of Open Architecture, pp. 6–9. Undersea Warfare, Washington (2008)

[351] Strachey, C.: Fundamental Concepts in Programming Languages. Higher Order Symbol. Comput. 13(1-2), 11–49 (2000)

[352] Stroustrup, B.: Adding Classes to the C Language: An Exercise in Language Evolution. Software: Practice and Experience 13(2), 139–161 (1983)

[353] Subramanian, K.R., Lee, S., Shiang, T.K., Sue, G.B.: Intelligent Agent Platform for Procurement. In: Preceedings of the IEEE International Conference on Systems, Man, and Cybernetics (IEEE SMC 1999), vol. 3, pp. 107–112 (1999)

[354] Suchman, L.A.: Plans and Situated Actions: The Problem of Human-machine Communication.Cambridge University Press, Cambridge (1987)

[355] Sundsted, T.: Agents on the Move. Java World (1998)

[356] Szarowicz, A., Francik, J., Mittmann, M., Remagnino, P.: Layering and heterogeneity as design principles for animated embedded agents. Journal of Information Sciences 171(4), 355–376 (2005)

[357] Szarowicz, A., Remagnino, P.: Avatars That Learn How to Behave. In: ECAI, pp. 554–558 (2004)

[358] Szyperski, C.: Component Software: Beyond Object-Oriented Programming, 2nd edn. Addison-Wesley, London (2002)

[359] Taghezout, N., Zaraté, P.: Supporting a multi-criterion decision making and multi-agent negotiation in manufacturing systems. Intelligent Decision Technologies 3(3), 139–155 (2009)

[360] Tambe, M.: Towards Flexible Teamwork. Journal of Artificial Intelligence Research 7, 83–124 (1997)

[361] Team, M.: The MadKit Project (a Multi-Agent Development Kit) (2002)

[362] Thagard, P.R.: Computational Philiosphy of Science. MIT Press, Cambridge (1993)

[363] Thomas, D., Armstrong-Helouvry, B.: Fuzzy logic control-a taxonomy of demonstrated benefits. Proceedings of the IEEE 83(3), 407–421 (1995)

[364] Thompson, A.: Silicon Evolution. In: Koza, J.R., Goldberg, D.E., Fogel, D.B., Riolo, R.L. (eds.) Proceedings of the First Annual Conference on Genetic Programming 1996, pp. 444–452. MIT Press, USA (1996)

[365] Thompson, K.: User's Reference to B. Case 39199-11 MM-72-1271-1-KT-pdp. Bell Laboratories, Murray Hill (1972)

[366] Toole, B.A.: Ada Byron, Lady Lovelace, An Analyst and Metaphysician. IEEE Annals of the History of Computing 18(3), 4–12 (1996)

[367] Touretzky, D.S., Pomerleau, D.A.: Reconstructing physical symbol systems. Cognitive Science: A Multidisciplinary Journal 18(2), 345–353 (1994)

[368] Tu, X., Terzopoulos, D.: Artificial Fishes: Physics, Locomotion, Perception, Behavior. In: Computer Graphics Annual Conference Series, vol. 28, pp. 43–50 (1994)

[369] Turing, A.: Intelligent Machinery. In: Meltzer, D. (ed.) Machine Intelligence, vol. 5, pp. 3–23. Edinburgh University Press (1948); National Physics Laboratory Report

[370] Turing, A.: Computing Machinery and Intelligence. Mind 59(236), 433–460 (1950) (unpublished until 1968)

[371] Tversky, A., Kahneman, D.: The framing of decisions and the psychology of choice. Science 211, 453–458 (1981)

[372] Tweedale, J.: Examining the Relevance of the Java Programming Language. Honors thesis, Monash University, Melbourne, Vic., Australia (1998)

[373] Tweedale, J., Bollebeck, F., Jain, L., Urlings, P.: Agent Transportation Layer Adaptation System. International Journal of Knowledge Based Intelligent Information and Engineering Systems (2007)

[374] Tweedale, J., Cutler, P.: Trust in Multi-Agent Systems. In: Gabrys, B., Howlett, J., Jain, L. (eds.) 10th International Conference on Knowledge Based Intelligent Information and Engineering Systems, pp. 479–485. Springer, Berlin (2006)

[375] Tweedale, J., Ichalkaranje, N., Sioutis, C., Jarvis, B., Consoli, A., Phillips-Wren, G.E.: Innovations in multi-agent systems. J. Network and Computer Applications 30(3), 1089–1115 (2007)

[376] Tweedale, J., Jain, L.C.: Interoperability with MAS. Journal of Intelligence and Fuzzy System 1(4), 175–181 (2007)

[377] Tweedale, J., Jain, L.C.: Multilingual Agents in a Dynamic Environment. Intelligent Decision Technologies (IDT) Journal, 193–202 (2008)

[378] Tweedale, J., Sioutis, C.: Mission Systems and Simulators. In: Challenges and Opportunities for a Complex and Networked World, Simulation Industry Association of Australia (SIAA), Linfield (2006)

[379] Urlings, P.: Intelligent Engineering Systems. In: Jain, L.C., Jain, R.K. (eds.) Second International Conference on Knowledge-Based Intelligent Engineering Systems. IEEE Press, USA (1998)

[380] Urlings, P.: Teaming Human and Machine: A conceptual framework for automation from an aeronautical perspective. Ph.D. thesis, University of South Australia, School of Electrical and Information Engineering (2004)

[381] Urlings, P., Brants, J., Zuidgeest, R., Eertink, B.: Crew Assistant: Architecture, EUCLID RTP 6.5 Crew Assistant. Technical Report DCAAWP1.3 NL, National Aerospace Laboratory, NLR, Amsterdam (1995)

[382] Urlings, P.J.M., Spijkervet, A.L.: Expert Systems for Decision Support in Military Aircraft. In: Murthy, T.K.S., Münch, R.E. (eds.) Computational Mechanics Publications, pp. 153–173. Springer, Berlin (1987)

[383] Valluri, A., Croson, D.: Agent learning in supplier selection models. Decision Support Systems 39, 219–240 (2005)

[384] Van Splunter, S., Wijingaards, N.J.E., Brazier, F.M.T.: Structuring Agents for Adaption. In: Alonso, E., Kudenko, D., Kazakov, D. (eds.) AAMAS 2003. LNCS (LNAI), vol. 2636, pp. 174–186. Springer, Heidelberg (2003)

[385] Vaucher, J., Ncho, A.: JADE Tutorial and Primer. Department d'informatique. Université de Montréal, Canada (2003)

[386] Vaux, J., Dale, R.: Review of Mind over Machine. AI & Society 1(1), 72–76 (1987)

[387] Venners, B.: Leading-Edge Java - How to Use Design Patterns: A Conversation with Erich Gamma. Artima Developer (2005), http://www.artima.com/lejava/articles/gammadpP.html (last Accessed March 29, 2007)

[388] Vincente, K.J.: Congnitive Work Analysis: Towards Safe, Productive and Healthy Computer-Based Work. Lawrence Erlbaum Associatesn, USA (1999)

[389] Von Neumann, J.: The computer and the brain. Yale University Press, USA (1958)

[390] Vonk, E., Jain, L.C., Johnson, R.P.: Automatic Generation of Neural Networks Architecture Using Evolutionary Computing. World Scientific Publishing Company, Singapore (1997)

[391] Wallis, P., Rönnquist, R., Jarvis, D., Lucas, A.: The Automated Wingman: Using JACK for Unmanned Autonomous Vehicles. In: IEEE Aerospace Conference Proceedings, vol. 5, pp. 2615–2622 (2002)

[392] Watkins, J.J.: Across the Board: The Mathematics of Chess Problems. Princeton University Press, Princeton (2004)

[393] Wellman, M.P., Durfee, E.H., Birmingham, W.P.: The Digital Library as Community of Information Agents. IEEE Expert Role of AI in Digital Libraries. 11(3), 10 (1996)

[394] White, G.: Java Command-Line Arguments. Dr. Dobb's Journal, Miller Freeman 21(2), 58–61 (1996)

[395] Whitehead, E., Murata, M.: XML Media Types. IETF, Networking Division of the USC Information Sciences Institute (ISI), CA (1998)

[396] Wicks, W.W.: Logic Design with Integrated Circuits. In: Venn Diagrams, pp. 36–49. Wiley, New York (1968)

[397] Wiener, E.L., Curry, R.: Flight-deck Automation: Promises and Problems. Technical Report NASA TM 81206, NASA, Moffett Field, CA (1982)

[398] Wilensky, R.: Planning and Understanding: A Computational Approach to Human Reasoning. In: Advanced Book Program, pp. 21–22. Addison-Wesley, Reading (1983)

[399] Wilke, J.: Wang drops workstation project. The Boston Globe, 3 Ian. Diery (1988)

[400] Williams, B.C., Nayak, P.P.: A Model-Based Approach to Reactive Self-Configuring Systems. In: Minker, J. (ed.) Workshop on Logic-Based Artificial Intelligence, June 14-16, 1999. Computer Science Department, University of Maryland, College Park, Maryland (1999)

[401] Winston, P.H.: Artificial Intelligence, 3rd edn. Addison-Wesley, Massachusetts (1992)

[402] Wirth, N.: Program Development by Stepwise Refinement. Communications of the ACM 14(4), 221–227 (1971)

[403] Wirth, N.: Hardware architectures for programming languages and programming languages for hardware architectures. SIGARCH Comput. Archit. News 15(5), 2–8 (1987)

[404] Wolfram, S., Weisstein, E.W.: Mathworld - Topology. Wolfram Media Inc, USA (2005)

[405] Wooldridge, M.: Verifiable semantics for agent communication languages. In: Proceedings of International Conference on Multi Agent Systems 1998, pp. 349–356 (1998)

[406] Wooldridge, M.: Verifying that agents implement a communication language. In: Joint Sixteenth National Conference on Artificial Intelligence (AAI 1999) and Eleventh Innovative Applications of Artificial Intelligence Conference (IAAI-1999), Orlando, FL, USA, pp. 52–57 (1999)

[407] Wooldridge, M.: An Introduction to MultiAgent Systems. John Wiley and Sons Ltd., Chichester (2002)

[408] Wooldridge, M., Jennings, N.: Agent Theories, Architectures, and Languages: A Survey. In: Wooldridge, M., Jennings, N.R. (eds.) Proceedings of ECAI 1994 Workshop on Agent Theories, Architectures, vol. 890, pp. 403–442. Springer, Heidelberg (1995)

[409] Wooldridge, M., Jennings, N.: Intelligent agents: theory and practice. Knowledge Engineering Review 10(2), 115–152 (1995)

[410] Wooldridge, M.J., Jennings, N.R.: Agent Theories, Architectures, and Languages: A Survey.In: Wooldridge, M.J., Jennings, N.R. (eds.): Intelligent Agents: ECAI 1994 Workshop on Agent Theories, Architectures, and Languages, pp. 1–39. Springer, Berlin (1994)

[411] Wooldridge, M., Jennings, N.R.: The Cooperative Problem-Solving Process. Journal of Logic and Computation 9, 563–592 (1999)

[412] Wooldridge, M., Muller, J., Tambe, M.: Agent theories, architectures, and languages: a bibliography. In: Proceedings, Intelligent Agents II. Agent Theories, Architectures, and Languages IJCAI 1995 Workshop (ATAL), pp. 408–431. IEEE Press, New York (1996)

[413] Yamakawa, T., Miki, T.: The Current Mode Fuzzy Logic Integrated Circuits Fabricated by the Standard CMOS Process. IEEE Transaction Computing 35(2), 161–167 (1986)

[414] Yamamoto, G., Nakamura, Y.: Architecture and Performance Evaluation of a Massive Multi-Agent System. In: Yamamoto, G., Nakamura, Y. (eds.) 3rd Annual Conference on Autonomous Agents, Seattle, Washington. ACM, New York (1999)

[415] Yato, T., Seta, T.: Complexity and Completeness of Finding Another Solution and Its Application to Puzzles. IEICE Transasactions of Fundamental Electronic Communication Computing and Science, pp. 1052–1060 (2003)

[416] Yeaton, T.: Black Duck Code Center. Black Duck, Waltham (2010)

[417] Yen, J., Fan, X., Sun, S., Hanratty, T., Dumer, J.: Agents with shared mental models for enhancing team decision makings. Decis. Support Syst. 41(3), 634–653 (2006)

[418] Yen, J., Yin, J., Ioerger, T.R., Miller, M.S., Xu, D., Volz, R.A.: CAST: Collaborative Agents for Simulating Teamwork. In: IJCAI, pp. 1135–1144 (2001)

[419] Zadeh, L.A.: Fuzzy Sets. Information and Control (1965)

[420] Zalewski, J.: Threads Primer: Review. IEEE Parallel & Distributed Technology (1996)

[421] Zelle, J.M., Mooney, R.J., Konvisser, J.B.: Combining Top-down and Bottom-up Techniques in Inductive Logic Programming. In: Eleventh International Conference of Machine Learning, pp. 341–351. Morgan Kaufmann, San Francisco (1994)

# Subject Index

# Authors

Dr. Jeffrey Tweedale is a Professional Scientific Engineering Officer working for the Avionics Mission Systems in Air Operations Division of System Sciences Laboratory within the Defence Science and Technology Organisation. He has been an adjunct member of the Knowledge-Based Intelligent Engineering Systems (KES) Centre since 2004. He has a PhD in Computer Systems Engineering from the University of South Australia, an MBA in Business from Adelaide University, B. Comp (Hons) degree in Computer Science from Monash University, a B.IT in Information Systems and a B. Ed from Melbourne University. His recent work interests include: Multi-Agent Systems, Unmanned Aerial Vehicles, System Automation, Human-Computer Trust, Electronic Engineering and Computer-Based Learning. His current research involves self-organising micro mission systems using airborne platforms and distributed sensors.

Jeffrey has over 120 research publications, with many as first author. He is a guest editor on several key Artificial Intelligence journals and has chaired many conference sessions. He has supervised over 20 students from the University of South Australia, the Central Queensland University and the Ritsumeikan University. He continues to adopt a mentoring role for colleagues and acquaintances. During his time at TAFE he instructed in Electronics Engineering and Communication. While at the University of South Australia, continues to instruct on Knowledge-Based Engineering and Digital Communications.

Professor Lakhmi C. Jain is a Director/Founder of the Knowledge-Based Intelligent Engineering Systems (KES) Centre, located in the University of South Australia. He is a fellow of the Institution of Engineers Australia.

His interests focus on the artificial intelligence paradigms and their applications in complex systems, art-science fusion, virtual systems, e-education, e-healthcare, unmanned air vehicles and intelligent agents.

9 783642 226755